# Food Engineering Series

Springer's *Food Engineering Series* is essential to the Food Engineering profession, providing exceptional texts in areas that are necessary for the understanding and development of this constantly evolving discipline. The titles are primarily reference-oriented, targeted to a wide audience including food, mechanical, chemical, and electrical engineers, as well as food scientists and technologists working in the food industry, academia, regulatory industry, or in the design of food manufacturing plants or specialized equipment.

More information about this series at http://www.springer.com/series/5996

Navneet Singh Deora
Aastha Deswal • Madhuresh Dwivedi
Editors

# Challenges and Potential Solutions in Gluten Free Product Development

Springer

*Editors*
Navneet Singh Deora
Chief Technology Officer
Bluetribe Foods
Mumbai, India

Aastha Deswal
Bright LifeCare Private Limited
Gurugram, Haryana, India

Madhuresh Dwivedi
Department of Food Process Engineering
National Institute of Technology
Rourkela, Odisha, India

ISSN 1571-0297
Food Engineering Series
ISBN 978-3-030-88696-7 ISBN 978-3-030-88697-4 (eBook)
https://doi.org/10.1007/978-3-030-88697-4



This Springer imprint is published by the registered company Springer Nature Switzerland AG
The registered company address is: Gewerbestrasse 11, 6330 Cham, Switzerland

# Preface

Recent epidemiological studies have shown that 1 in 100 people worldwide suffer from gluten intolerance. The worldwide average of celiac sufferers has been predicted to increase in the next number of years. This will result in a rapid-growing market for high-quality gluten-free cereal products. However, due to the unique properties of gluten, it is a major challenge for food scientists and industries to manufacturer quality gluten-free products at present.

Additionally, gluten absence results in major complications for product developers. Currently, many gluten-free products available in the market are of low quality and limited shelf life, and exhibit inferior mouthfeel and overall acceptability as compared to the products containing gluten. This presents a major challenge to the cereal technologist and baker alike, and has led to the search for alternatives to gluten in the manufacture of gluten-free bakery products. This book aims to provide the possible solutions for the gluten-free product development.

The only treatment for celiac disease is the total lifelong avoidance of gluten ingestion. Patients have to follow a very strict diet and avoid any products that contain wheat, rye, or barley (some authors also include oats). Avoidance of these cereals leads to a recovery from the disease and significant improvement of the intestinal mucosa and its absorptive functions. Patients with celiac disease cannot eat some common foods such as bread, pizzas, and biscuits or drink beer. This book also aims to cover the nutrition aspects of gluten-free products. Since protein is of paramount importance, this book will also serve to include the alternative protein in gluten-free product development in form of case study.

The overall aim of this book is to provide the reader a chance to take a journey through all aspects related to product development for patients suffering from celiac disease. As such, this book is unique in its form and hopes to represent a technical guide to the readers working in the related areas. It aims to summarize and critically review the works and knowledge gained so far in the area of gluten-free product development.

Mumbai, India — Navneet Singh Deora
Gurugram, Haryana, India — Aastha Deswal
Rourkela, Odisha, India — Madhuresh Dwivedi

# Acknowledgment

This book is a result of the combined efforts of highly qualified scientists. Each contributor was responsible for researching and reviewing subjects of immense depth, breadth, and complexity. Care and attention were paramount to ensure technical accuracy for each chapter discussed in this book. This book is unique as stated earlier, and it is our sincere hope and belief that it will serve as an essential reference for gluten-free products.

We wish to thank all the contributors for sharing their expertise throughout our journey. We also thank the reviewers for giving their valuable comments, leading to improvements in the content of each chapter.

The production of this current book could not have been accomplished without the hard work and excellent suggestions of professionals in the production team assigned by Springer to manage the project.

# Disclaimer

All information in this book is based on practical knowledge gained by the author while working in factories as well as theoretical knowledge gained during his studies and should not be used as the basis for any legal claims. Hence, all information stated is not intended to credit or discredit any manufacturer of equipment or additives and is based purely on the opinion of the authors.

# Contents

# Contributors


**R. Anand Kumar** Department of Food Process Engineering, NIT Rourkela, Rourkela, Odisha, India

**Saptashish Deb** Department of Food Engineering and Technology, Sant Longowal Institute of Engineering & Technology, Sangrur, Punjab, India

Center for Rural Development and Technology, Indian Institute of Technology Delhi, New Delhi, India

**Madhuresh Dwivedi** Department of Food Process Engineering, NIT Rourkela, Rourkela, Odisha, India

**Payel Ghosh** Department of Food Technology, Vignan's Foundation for Science Technology and Research, Vadlamuḍi, Andhra Pradesh, India

**Rewa Kulshrestha** Department of Food Processing and Technology, Atal Bihari Vajpayee Vishwavidyalaya, Bilaspur, Chhattisgarh, India

**Chitrangada Das Mukhopadhyay** Centre for Healthcare Science and Technology, Indian Institute of Engineering Science and Technology, Shibpur, Howrah, West Bengal, India

**Mohona Munshi** Department of Food Engineering and Technology, Sant Longowal Institute of Engineering & Technology, Sangrur, Punjab, India

Department of Chemical Engineering, Vignan Foundation for Science, Technology, and Research, Vadlamudi, Guntur, Andhra Pradesh, India

**Surabhi Pandey** Bioresource Engineering Department, McGill University, Ste-Anne-de-Bellevue, Canada

**Sumit Sudhir Pathaka** Department of Food Process Engineering, National Institute of Technology Rourkela, Rourkela, Odisha, India

**Rama Chandra Pradhana** Department of Food Process Engineering, National Institute of Technology Rourkela, Rourkela, Odisha, India

**Sandeep Singh Rana** Department of Food Technology, Vignan's Foundation for Science, Technology and Research, Vadlamudi, Andhra Pradesh, India

**E. J. Rifna** Department of Food Process Engineering, NIT Rourkela, Rourkela, Odisha, India

**Winny Routray** Department of Food Process Engineering, NIT Rourkela, Rourkela, Odisha, India

**Murakonda Sahithi** Department of Food Process Engineering, NIT Rourkela, Rourkela, Odisha, India

**Mahipal Singh Tomara** Department of Food Process Engineering, National Institute of Technology Rourkela, Rourkela, Odisha, India

# About the Editors

**Navneet Singh Deora** is a food engineer with a PhD from the Agricultural & Food Engineering Department, Indian Institute of Technology (IIT) Kharagpur, India; has a master's degree in food technology from CSIR-CFTRI, Mysore, India; and is a Bachelor of Engineering in Food Processing Technology from AD Patel Institute of Technology (ADIT), Gujarat. In the past, he was involved with RPSG Goenka Group (*Extrusion Expert*, 2019), Jubilant Food Works (*Open Innovation*, 2018), and Nestle R&D (*Cereal Specialist*, 2017). His research interests focus on open innovation in the food industry, sustainability, plant-based proteins, equipment design, and advancements in cereal science. He is an author or co-author of multiple research papers, books, and patents. He is currently chief technology officer of Bluetribe Foods.

**Aastha Deswal** is a PhD from the Agricultural & Food Engineering Department, Indian Institute of Technology (IIT) Kharagpur, and graduate from CFTRI, Mysore. She has over 10 years of research experience in the field of food technology at both academic and industrial levels covering various topics but mostly focused on product development. She has hands-on experience of working with gluten-free products in the past, which has been very helpful while editing this book. She has authored many research articles, chapters, and books covering various topics of food technology.

**Madhuresh Dwivedi,** MTech, PhD, is an assistant professor in the Department of Food Process Engineering at the National Institute of Technology (NIT) Rourkela. Dr. Dwivedi obtained his BTech degree in agricultural engineering from the College of Agricultural Engineering, Jabalpur India, in the year 2010, MTech in food process engineering from the Indian Institute of Technology (IIT) Kharagpur in the year 2012, and PhD (2015) from the Department of Food Process Engineering, Indian Institute of Technology (IIT) Kharagpur.

# Chapter 1
# Current Advances in Celiac Disease: Consequences and Improvement Strategies

**Chitrangada Das Mukhopadhyay**

**Abstract** Celiac is a chronic enteric disease resulted from an abnormal immune response to gluten proteins in patients having a certain genomic constitution. Tissue transglutaminase enzyme 2 converts the glutamine residues of gluten peptides by deamidation reaction into glutamic acid, which binds to human leukocyte antigen (HLA)-DQ2 or -DQ8 molecules and subsequently evokes T cell responses leading to small intestine inflammation. These events lead to the typical symptoms associated with celiac disease. Also, wheat proteins are rich in proline content and are resistant to human pancreatic and gastric enzymes. Different peptidases from microbial and fungal sources can degrade these incompletely digested peptides. While following a gluten-free diet is the best preventive strategy, a combination therapy by using proteases or carboxypeptidases from microbial sources for gluten detoxification or treatment with tissue transglutaminase inhibitors may also be a good option. Intestinal epithelial cell lines (Caco-2) may be used as *in vitro* model to study trans/paracellular permeability, distortion of intercellular tight junction protein viz., occludin, and ZO-1, and rearrangement of actin filaments.

**Keywords** Celiac · Tissue transglutaminase · Gluten · Enzyme therapy

## 1.1 Introduction

Celiac disease is an autoimmune disease caused by an intolerance to gluten proteins (Dewar et al., 2003; Simón et al., 2017; Lo et al., 2003; Cellier et al., 2000; Clemente et al., 2003). This disease develops in genetically predisposed persons, who upon ingestion of gluten-rich food such as products of wheat, rye, barley, etc (Lo et al., 2003). The pathological symptoms include a flattened mucosal layer of small intestinal with uncontrolled epithelial cell proliferation, presence of lymphocytes,

C. D. Mukhopadhyay (✉)
Centre for Healthcare Science and Technology, Indian Institute of Engineering Science and Technology, Shibpur, Howrah, West Bengal, India



N. Singh Deora et al. (eds.), *Challenges and Potential Solutions in Gluten Free Product Development*, Food Engineering Series,
https://doi.org/10.1007/978-3-030-88697-4_1

cryptic hyperplasia, and lesser differentiation of enterocytes (Cellier et al., 2000). This leads to impaired absorptive function of the small intestine. Gluten contains a very high amount of proline (15%) and glutamine (35%) . In celiac patients, this gluten is resistant to normal luminal digestion and becomes toxic to the intestine. The antigenic property of some epitopes on gluten and their role in disease progression has been identified. Any structural alteration to this antigenic potential of the toxic protein may lead to a novel therapeutic pathway. It is now established that induction of the deamidating activity of tissue transglutaminase (tTG) triggers the pathogenic immune responses. Gliadin peptide, a part of gluten protein is a good substrate of tTG. So, enzymatic degradation of the antigenic epitope in toxic peptides coupled with inhibition of tTG activity may become an alternative treatment strategy for CD patients.

Many of the patients with celiac disease are asymptomatic or have mild symptoms. With the advanced diagnostic systems more silent cases are reported which shows that the prevalence of this disease is much more than thought earlier. This autoimmune disease is highly prevalent in Western countries (1% of the total population in European countries) but also diagnosed to be a serious disorder and increasing importance in Northern India. A stringent gluten-free diet (GFD) can help to overcome this enteropathy, however cross-contamination, improper labeling contributes to the immune response in patients. An oral dietary supplement may restore immunological tolerance and represent the ideal cure for celiac disease. In this connection enzymatic hydrolysis of α- gliadin of gluten by proteolytic and amylolytic enzymes coupled with small-molecule inhibition of tTG should be an efficient therapeutic approach (Lorand & Graham, 2003). In this chapter, an overview of the disease, its pathological complexities, detection method, prevention, and therapeutic interventions have been discussed in detail.

## 1.2 Overview of the Disease

Microscopic analysis of the sections of the inner mucosal layer of human small intestinal from a normal person shows several villi with one layer of columnar epithelial cells having basal nuclei, intraepithelial lymphocytes (IEL), and plasma cells in the lamina propria of the small intestine generally with villous to crypt ratio of 1:5.

On the contrary, the sections of the specimen from the CD patients are characterized by complete loss of villi, plane mucosal layer, having ridges and of the crypt openings onto the luminal surface, appearance of abnormal squamous epithelial cells, an increase in the number of lymphocytes and plasma cells infiltrated in the intraepithelial area and lamina propria, increase in mitoses of crypts. Loss of villi causes decreased surface area for absorption of nutrients in the small intestine and patients to suffer from diarrhea, severe weight loss, iron deficiency led anemia, malnutrition (Clemente et al., 2003; Molberg et al., 2003; Shewry et al., 1992; Lorand & Graham, 2003; Solid, 2000).

## 1.3 Gluten and Gluten-Free Diet (GFD)

Wheat contains 8–17% protein based on its breeding variety and environmental makeup. The water-insoluble protein fraction of wheat protein constitutes a visco-elastic mass, called gluten. Gluten is a complex protein made up of polymeric and monomeric subunits and constitutes approximately 78–85% of the total wheat protein rich in proline and glutamine. Gluten is responsible for water absorption capacity, baking quality, viscosity, cohesiveness, and elasticity of the dough (Han et al., 2013). Gluten is classified into two main fractions viz., gliadin, which is soluble in aqueous alcohol, and glutenins which are insoluble in the same. Gliadins are monomeric proteins (28–55 KDa) and further subdivided as α/β-, γ- and ω-type based on their primary structure. Glutenins have high (67–88 KDa) or low molecular weight (32–35 KDa) peptides connected by intermolecular disulfide bonds (Khosla, 2017). Gluten proteins are also strongly connected by hydrogen bonds, ionic bonds, and hydrophobic bonds which confer structural and physical properties of the gluten. Glutenins provide elasticity, and gliadins provide viscosity to the gluten complex (Deora, 2017; Daum et al., 1999; Moss et al. 1996; Schuppan et al., 2009; Han et al., 2013). The gliadins having high glutamine and proline content and humans lack endopeptidases to cleave intermolecular bonds between them. The incompletely digested gliadin are immunogenic to CD patients.

## 1.4 Clinical Symptoms and Possible Mechanism of Villous Atrophy

CD patients manifest mild to severe intestinal inflammation characterized by an increase in intraepithelial lymphocytes, infiltration of mononuclear cells from the subepithelial layer, increase in epithelial mitosis causing total villous atrophy and crypt hyperplasia. Loss of villi is the histopathologic hallmark of celiac disease. This may be due to degradation of intracellular tight junction proteins, caspase-mediated apoptosis, and defect in epithelial regeneration (Kelly et al. 2015; Butterworth & Louis, 2019; De et al., 2017; Palejwala & Watson, 2000). Briefly, the sequence of events starts with the ingestion of gluten proteins by the persons having impaired genetic makeup. HLA-DQ2 and DQ8 heterodimers react with incompletely digested gluten peptides and are presented to CD4+ T cells in the small intestine (Ciccocioppo et al., 2002; Miura et al., 2005; Marsh & Crowe, 1995; Lionetti & Catassi, 2011; Gujral et al. 2012; Kelly et al. 2015). The incompletely digested gluten peptides cross the epithelium by conventional trans paracellular transport and reach the subepithelial region. Activation of macrophages and dendritic cells increased production of IFN I and II and several cytokines in the small intestine subsequently initiates humoral immunity. The release of matrix metalloproteinases can also damage the intestinal mucosa.

## 1.5 Factors Causing Celiac Disease

### *1.5.1 Environmental Factors*

This disease is stimulated by the consumption of gluten – a protein found in wheat endosperm- or of related proteins found in other grains like rye and barley (Fernandez-Jimenez et al., 2014; Hollon et al., 2015). Abundant gluten protein acts as an environmental trigger. Removal of gluten from daily diet prevents the autoimmune process and again may be restored by the slightest consumption of gluten. Intestinal tight junctions come apart and allow a large amount of incompletely digested gluten into the small intestine. An increase in tissue transglutaminase (tTG) activity acts as an autoantigen. Oats, rice, maize, sorghum, and millet do not activate the celiac disease. The high proline-containing proteins escape proteolytic digestion in the human intestine (Gujral et al., 2012; Moss et al. 1996; Arzu, 2010; Dickson et al., 2006).

### *1.5.2 Genetic Factors and HLA Class II Genes*

CD has a strong association with HLA class II genes on DQ locus (Karell et al., 2003) for the phenotypic expression of disease. HLA-DQ2 or -DQ8 heterodimers are relatively common in white populations. Persons homozygous for DR3 results only DQ2 molecules; but if they are DR3(DR17)/17 heterozygous then 50% of the DQ molecules are DQ2, and, in those who are heterozygous for DR5/7 or DR3 (DR17)/ other, 25% of the DQ molecules are DQ2 (Margaritte et al., 2004; Ehrmann et al., 2003; Augustin et al., 2005). Celiac disease is prevalent in DR3 homozygous or DR3/17 heterozygous genetic makeup.

### *1.5.3 Binding of DQ2 to Gluten Peptides*

DQ2 molecules have active sites for –ve charges residues. They favor left-handed poly-proline II helical configuration, similar to that of gliadins. It has been estimated that there are approximately 50 aa in wheat, 60 aa in the rye, and 35 aa in barley that have the potential to bind with DQ2 or DQ8 and considered to be the candidate sequences for activating celiac disease (Kim et al., 2004).

## 1.6 Importance of tTG

Human (tTG) is an allosterically regulated enzyme involved in cell signaling blood clotting, wound healing, extracellular matrix formation, and cell adhesion and motility. tTG causes sequence-specific deamidation of dietary gluten peptides in the small intestines of CD patients (Marsh & Crowe, 1995). They are found in intracellular as well as extracellular environments of many organs. The catalytically active form of tTG2 or "open" form can crosslink specific g-glutaminyl and 3-lysine residues on proteins in the presence of calcium and absence of GTP or GDP and forms proteolytically resistant covalent bonds. In the reverse condition, TG2 assumes an inactive/ "closed" form. Human tTG also possesses $Ca^{2+}$-binding sites, N-terminal fibronectin association site, and guanosine triphosphate hydrolysis site. In celiac sprue, peptides derived from dietary gluten are deamidated by TG2 and enhances their affinity towards MHC, and induces humoral immunity. Inhibition of TG2 activity may be a suitable target for CD therapy.

### *1.6.1 Modification of Gluten by tTG*

The presence of serum antibodies to tissue transglutaminase (tTG) in CD patients confirms its autoimmune response. tTG is the main autoantigen among the antibodies produced by gluten antigen called endomysial antibodies (EMA) (Dieterich et al., 1997). tTG favors the formation of isopeptide bonds between the γ-carboxamide group of glutamine residue of a peptide to the ε-amino group of a lysine residue or other biogenic molecules viz., putrescine, spermidine, spermine, and histamine in low pH. Gluten acts as an exogenous trigger for the production of tTG-specific autoantibodies in CD patients and also generates other antigenic epitopes by cross-linking of gliadin peptides to itself and/or to other peptides leading to autoimmunity.

### *1.6.2 Role of Intraepithelial Lymphocytes and Muscular Pericytes*

An increase in intraepithelial lymphocytes is a characteristic feature of celiac disease. Interleukin (IL)-15 plays a major role and is up-regulated by epithelial cells and cells in the lamina propria in CD patients (Karell et al., 2003; Margaritte-Jeannin et al., 2004; Myrsky et al., 2008; Naluai et al., 2001; Ploski et al., 1993; Louka et al., 2002 ; Mazzarella et al., 2003; Kim et al., 2004; Singh et al., 1995; Liu et al., 2002).

### *1.6.3 Auto Antibody-Mediated Disease Pathogenesis*

The gluten-induced disease-specific autoantibodies might constitute in CD pathogenesis. Angiogenesis commences with the formation of the endothelial tube (Kale et al., 2005; Lorand, 2007). Then mesenchymal cells accumulate surrounding the endothelial tubes and later differentiates into vascular smooth muscle cells or pericytes. Coeliac disease-specific autoantibodies hinder angiogenesis. CD-specific autoantibodies inhibit angiogenesis and disorganize the actin cytoskeleton. Cytoskeletal disarrangement inhibits cell migration, thus the entire vasculature network seen in the small-intestinal mucosa shows abnormality. Overexpression of TG2 in cells leads to reduced cell migration. The autoantibody deposits have been identified around the blood vascular system of the liver causing mild liver in CD patients. Auto-antibodies in the blood vessels of the brain of CD patients suffer from neurological problems and neuroblast apoptosis (Cervio et al., 2007; Pinkas et al., 2007; Lai et al., 2010; Lai et al., 2008; Andringa et al., 2004)

## 1.7 Detection of Celiac Disease

The diagnosis of CD is very important because delayed diagnosis adversely affects the health and quality of life of patients. The diagnosis of CD requires accurate, sensitive, simple, specific, and cheaper as well-as non-invasive analytical tools (Sollid & Lundin, 2009; Stamnaes et al., 2010; Dieterich et al., 1997; Farre, 2014). CD-specific autoantibody detection in patient serum and saliva may be done using electrochemical, optic-fiber, piezoelectric biosensors, and POC finger-prick methods. CD-specific volatile organic compounds (VOCs) in urine and faeces may also be used as a sample for detection. The most common methods of celiac disease detection are discussed here.

Blood tests are considered the most important. This includes detection of total IgA, tTG IgA, and EMA IgA at first. A biopsy of the small intestine can confirm the findings of the blood test. This is done by endoscopy of the small intestine which is another most reliable detection method. Because the symptoms of celiac disease can be varied, it is often undiagnosed or misdiagnosed. Detection of CD-specific antibodies in blood serum may require additional testing, preferably a DNA test for an accurate diagnosis. Genetic testing to check the presence of DQ2 or DQ8 genes are necessary to diagnose CD.

## 1.8 Prevention of Celiac Disease

### *1.8.1 Adherence to GFD*

The best option to get rid of CD is adherence to a gluten-free diet life-long. However complete removal of gluten from the daily diet is difficult. Several other grains like maize and rice, soya beans, and starch sources are recommended for a gluten-free diet. Some other options for GFD include tapioca, sorghum, carob, buckwheat, and millet. The main problem related to a gluten-free diet is the poor taste of gluten-free products. Additionally, improper food labels, cross-contamination, and lack of awareness are also related to low compliance. GFD draws social and financial limitations creating a huge impact on family life, workplace. travelling etc (Lerner et al., 2017; Donaldson et al., 2015; Farre, 2014, Guandalini and Rose, 2012, Lequin, 2005). Also, immunological issues are not influenced by diet. The use of a "natural gluten-free diet" offers the greatest compliance and the lowest risk of nutritional imbalance.

### *1.8.2 Problems in Maintaining a Strict GFD*

Gluten is extensively used in the food industry because of its unique viscoelastic properties discussed earlier in this chapter. CD patients are challenged with several issues like insufficient information about the disease, food contamination, and improper food labeling on the packaged food items (Myrsky et al., 2008; Brusca, 2015; Guandalini & Rose, 2012; Hall et al., 2009). Gluten is also present in ice-creams, sweets, confectionary foods, spreads and seasonings, beer, soups and sauces, malted beverages, and many more which need to be avoided. The contamination of food with gluten can occur: during milling, preparation of commercial food products, harvesting, storage, and packaging of grain bags by the farmers (Cervio et al., 2007; Farre, 2017; Lerner et al., 2017). A team approach, including patients, family, physicians, and dietician is required to manage CD patients. After the diagnosis is made, patients should go for nutritional assessment, meal planning, diet education, and counseling with the social and emotional adaptation to the GF lifestyle (García-Manzanares & Lucendo, 2011; Saturni et al., 2010; See & Murray, 2006).

### *1.8.3 Non-availability of GF (Gluten-Free) Food Products*

The number of patients with CD in India is low, thus the commercial production of GFD is very limited. Production of mainly flour and biscuits are being produced but quality control is not done (Green & Cellier, 2007; Di Sabatino & Corazza, 2009;

Zarkadas et al., 2006; Lee et al., 2007). Non-availability of GFD outside their home environment restricts their travel, occupation, and profession. GF food items are considerably more expensive than regular gluten-containing food (J.G. Donaldson, 2015; Butler, 2015; Lequin, 2005; Guandalini et al., 2005; Olsson et al., 2008).

## 1.9 Counselling of the CD Patients Undergoing GFD

### *1.9.1 Patient Education and Awareness*

The management of CD patients involves education of the patients and their families about the disease and dietary restrictions, timely visits, consistent supervision, and guidance of the nutritionist are important. Also, the disease status should be monitored regularly (Hall et al., 2009; García-Manzanares & Lucendo, 2011; Saturni et al., 2010; Stevens & Rashid, 2008).

### *1.9.2 CD Management and Counselling by Dieticians*

The dietary counselor should have sufficient knowledge about GF food and food products. Prescribing GFD and a specific well-balanced diet is necessary (Stevens & Rashid, 2008; Olsson et al., 2008). The dietician is the most qualified health care professional to provide nutrition therapy. They have extensive academic and practical experience including in-depth knowledge of nutrition, nutritional needs, nutrition composition, and food preparation information, and educational factors that affect the food and nutrition behavior of people (Nasr et al., 2012; Niewinski, 2008).

### *1.9.3 Celiac Disease Support Groups*

Celiac disease support groups provide a platform for patients to discuss their problems amongst themselves and learn from each other, provides information about GF products and their availability, but also can act as an advocacy for gluten labeling and other issues to the Government and regulatory bodies (Green & Cellier, 2007; Di Sabatino & Corazza, 2009; Zarkadas et al., 2006; Lee et al., 2007). Furthermore, it has been observed that the adherence to a GFD increases when individuals are members of a patient support group (Addolorato et al., 2004; Silvester & Rashid, 2010; Leffler et al., 2008 ; Catassi et al., 2007).

### 1.9.4 Development of Reliable GFD at a Large Scale

There is a need for large-scale industrial-level production of reliable and affordable GF food, including choices of food products ranging from snacks, flour, sweets, ice-creams, and ready-to-eat packets. All GF food products should undergo a quality check (Lanzini et al., 2009; Akobeng & Thomas, 2008; Haush et al., 2002; Shan et al., 2002). It is therefore essential that food products, which are available, should be labeled for gluten content.

### 1.9.5 Therapeutic Strategies

Different therapeutic strategies and ongoing commercial interventions are discussed below.

### 1.9.6 Enzyme Therapy

Incomplete digestion of gluten protein may be digested by some endopeptidases of microbial origin. Prolyl endopeptidases (PEP), are such enzymes having endoproteolytic activity. A two-enzyme cocktail comprised of a glutamine-specific cysteine protease (EP-B2) that functions under gastric conditions and a PEP, which acts simultaneously with pancreatic proteases in duodenal microenvironments, is a potent candidate for celiac sprue therapy (Piper et al., 2004; Marti et al., 2005). Cell culture-based in vitro analyses, in vivo preclinical assessment, and ex vivo (human) experimental approaches have confirmed that PEPs can readily cleave proline-rich gluten peptides, thereby reducing the antigenic burden of gluten (Pyle et al., 2005; Gass et al., 2005; Wood et al., 2020). Further cleavage at the intestinal surface by brush-border aminopeptidases and carboxypeptidases may be helpful. The possibility to formulate an oral enzyme treatment for celiac sprue using PEPs is under consideration. Enteric-coated pills protect the enzymatic activity from the harsh conditions of the stomach and rapidly release its contents under simulated duodenal conditions (Tripathi et al., 2009; Paterson et al., 2007; Molberg et al., 2000; Porta et al., 1990; Esposito et al., 2003)

### 1.9.7 Zonulin Antagonists

Zonulin is a 47 KDa protein, overexpressed in intestinal tissue of patients with Cd compared to healthy normal (Rashtak et al., 2011). This membrane receptor could induce an increase in intestinal permeability through a tight junction (TJ) re-arrangement (Xia et al., 2007; Kapoerchan et al., 2008; Beaurepaire et al., 2009; Yokoyama et al., 2009) . The exposure of intestinal epithelium to gliadin in CD patients causes zonulin release from enterocytes through recruitment of MYD-88. Mucosal CXCR3 expression also increases in active CD but returns to baseline levels following a gluten-free diet. Gliadin induces a physical association between CXCR3 and MyD88 in enterocytes (Molberg et al., 2003). AT-1001 is an octapeptide inhibitor of paracellular permeability that acts as a competitive agonist of zonulin. AT-1001 has reached phase 2 clinical trial where efficacy and safety are being tested (Rossi et al., 2012).

### *1.9.8 TTG Inhibitor*

As discussed in previous paragraphs that human tTG-2 plays a very significant role in CD pathology. Thus inhibition of tTG activity by specific small molecule inhibitors may represent a good therapeutic option . The following criteria should be considered to synthesize tTG inhibitor.

- tTG activity can be inhibited by developing specific inhibitors targeting the active site. Inhibitors should be safe and effective.
- An inhibition should limit excessive tTG activity that promotes CD and antagonizes tTG cross-linking reaction.
- Lys residues are critical for enzyme function, solubility, protein-protein interactions. Alteration of free Lys-residues may hinder the cross-linking reaction.
- Fluorescent tagging of inhibitor molecule will help in *in-vivo* image analyses.

### *1.9.9 DQ2/DQ8 Inhibitor*

The MHC protein HLA-DQ2 and DQ8 are two very important pharmacological targets (Xia et al., 2007; Kapoerchan et al., 2008). Siegel et al designed an aldehyde-bearing gluten peptide analogue to bind HLA-DQ2 ligand and a reversible tTG 2 inhibitor based on the HLA-DQ2 crystalline structure.

### *1.9.10 Other Therapeutic Options*

Monoclonal antibodies to proinflammatory cytokines are used for the treatment of CD patients. They respond to gluten intolerance immediately after gluten intake and releases IFNγ and IL-15, and other cytokines . An *in vitro* study showed that the inhibition of interleukin- 15 might have the potential to control CD.

In Australia, a vaccine for CD was developed which includes peptides accounting for T-cell activation to gluten in patients with HLA DQ2-associated CD. These peptides have been converted into a pharmaceutical formulation capable of inducing immune tolerance in a rodent model (Di Sabatino et al. 2018).

## 1.10 Indian Scenario of Celiac Disease: Problems and Challenges

CD emerging in some parts of North India is becoming a major health issue. Many of the affected persons are not diagnosed. More awareness among healthcare professionals and caregivers as well as the general public, accurate and cheap method of detection may help the patients to proper treatment and counseling. There are a number of issues, which require urgent attention. Team-based patient management proper supervision of patients, training of nutritionists, large scale industrial production of reliable and affordable GFD, detailed labeling of packaged foods about the nutritional value and gluten content, variety of palatable gluten-free food may ease the daily life of CD patients.

## 1.11 Conclusion

GFD is the common control strategy to control CD. However lifelong maintenance of a strict diet is a difficult task since gluten is the most common ingredient in the human diet. The development of a dietary supplement composed of enzymes from natural sources can lead to an alternative treatment strategy for celiac patients. Regression of gluten antigenicity and inhibition of tTG mediated crosslinking and deamidation of gluten epitopes and thus subsequent prevention of gluten interaction with HLA restricted T cells can help to manage CD. This might inhibit gluten-specific T-cell response in the small intestine. In celiac patients, tTG reacts with Gln residues in gliadin to form tTG-gliadin complexes. An enzyme-based therapeutic strategy has the potential to improve the quality of life of CD patients.

## References

Addolorato, G., de Lorenzi, G., Abenavoli, L., Leggio, L., Capristo, E., & Gasbarrini, G. (2004). Psychological support counselling improves gluten-free diet compliance in coeliac patients with affective disorders. *Alimentary Pharmacology & Therapeutics*, 777–782.

Akobeng, A. K., & Thomas, A. G. (2008). Systematic review: Tolerable amount of gluten for people with coeliac disease. *Alimentary Pharmacology & Therapeutics, 27*, 1044–1052.

Andringa, G., Lam, K. Y., Chegary, M., Wang, X., Chase, T. N., & Bennett, M. C. (2004). Tissue transglutaminase catalyzes the formation of alpha-synuclein crosslinks in Parkinson's disease. *FASEB Journal*, 932–934.

Augustin, M. T., Kokkonen, J., & Karttunen, T. J. (2005). Evidence for increased apoptosis of duodenal intraepithelial lymphocytes in cow's milk sensitive enteropathy. *Journal of Pediatriac Gastroenterology Nutrition, 40*(3), 352–358.

Beaurepaire, C., Smyth, D., & McKay, D. M. (2009). Interferon-gamma regulation of intestinal epithelial permeability. *Journal of Interferon Cytokine Research, 29*(3), 133–144.

Brusca. (2015). Overview of biomarkers for diagnosis and monitoring of celiac disease. *Advances in Clinical Chemistry, 68*, 1–55.

Butterworth, J. R., & Louis, L. (2019). Coeliac disease. *Medicine, 47*(5), 314–319.

Catassi, C., Fabiani, E., Iacono, G., D'Agate, C., Francavilla, R., Biagi, F., Volta, U., Accomando, S., Picarelli, A., de Vitis, I., et al. (2007). A prospective, double-blind, placebo-controlled trial to establish a safe gluten threshold for patients with celiac disease. *American Journal of Clinical Nutrition*, 160–166.

Cellier, C., Delabesse, E., Helmer, C., Patey, N., Matuchansky, C., Jabri, B., Macintyre, E., Cerf-Bensussan, N., & Brousse, N. (2000). Refractory sprue, coeliac disease, and enteropathy-associated T-cell lymphoma. *Lancet*, 203–208.

Cervio, E., Volta, U., Verri, M., et al. (2007). Sera of patients with celiac disease and neurologic disorders evoke a mitochondrial-dependent apoptosis in vitro. *Gastroenterology, 133*(1), 195–206.

Ciccocioppo, R., Sabatino, A. D., et al. (2002). Mechanisms of villous atrophy in autoimmune enteropathy and coeliac disease. *Clinical and Experimental Immunology*, 88–93.

Clemente, M. G., De Virgiliis, S., Kang, J. S., Macatagney, R., Musu, M. P., Di Pierro, M. R., Drago, S., Congia, M., & Fasano, A. (2003). Early effects of gliadin on enterocyte intracellular signaling involved in intestinal barrier function. *Gut, 52*, 218–223.

Daum, S., Bauer, U., Foss, H. D., Schuppan, D., Stein, H., Riecken, E. O., & Ullrich, R. (1999). Increased expression of mRNA for matrix metalloproteinases-1 and -3 and tissue inhibitor of metalloproteinase-1 in intestinal biopsy specimens from patients with coeliac disease. *Gutology, 44*, 17–25.

De Re, V., Magris, R., & Cannizzaro, R. (2017). New insights into the pathogenesis of celiac disease. *Frontiers in Medicine, 4*, 137.

Deora, N. S. (2017). Gluten free detection – A recent insight EC gastroenterology and digestive. *System*, 148–151.

Dewar, D., Pereira, S. P., & Ciclitira, J. (2003). The pathogenesis of Celiac disease. *International Journal of Biochemistry*, 17–24.

Di Sabatino, A., & Corazza, G. R. (2009). Coeliac disease. *Lancet*, 1480–1493.

Di Sabatino, A., Lenti, M. V., Corazza, G. R., Gianfrani, C. (2018). Vaccine Immunotherapy for Celiac Disease. Front Med (Lausanne). 5:187. https://doi.org/10.3389/fmed.2018.00187

Dickson, B. C., Streutker, C. J., & Chetty, R. (2006). Coeliac disease: An update for pathologists. *Journal of Clinical Pathology, 59*(10), 1008–1016.

Dieterich, W., Ehnis, T., Bauer, M., Donner, P., Volta, U., Riecken, E. O., & Schuppan, D. (1997). Identification of tissue transglutaminase as the autoantigen of celiac disease. *Nature Medicine*, 797–801.

Donaldson, J. G. (2015). Immunofluorescence staining. *Current Protocol in Cellular Biology*, 4–7.

Ehrmann, J., Kolek, A., Kodousek, R., et al. (2003). Immunohistochemical study of the apoptotic mechanisms in the intestinal mucosa during children's coeliac disease. *Virchows Archive, 442*(5), 453–461.

Arzu, E. (2010). Gluten-sensitive enteropathy (celiac disease): Controversies in diagnosis and classification. *Archives of Patholology & Laboratory Medicine, 134*(6), 826–836.

Esposito, C., Paparo, F., Caputo, I., et al. (2003). Expression and enzymatic activity of small intestinal tissue transglutaminase in celiac disease. *American Journal of Gastroenterology*, 1813–1820.

Farre, C. (2014). The role of serology in celiac disease screening, diagnosis and follow-up. In: L. Rodrigo, & A. S. Pena (Eds.), *Celiac disease and non-celiac gluten sensitivity* (pp. 151–169). OmniaScience.

Fernandez-Jimenez, N., Plaza-Izurieta, L., & Bilbao, J. R. (2014). Genetic markers in celiac disease. In L. Rodrigo & A. S. Pena (Eds.), *Celiac disease and non-celiac gluten sensitivity* (pp. 103–121). Omnia Science.

García-Manzanares, A., & Lucendo, A. J. (2011). Nutritional and dietary aspects of celiac disease. *Nutritional Clinical Practice*, 163–173.

Gass, J., Ehren, J., Strohmeier, G., Isaacs, I., & Khosla, C. (2005). Fermentation, purification, formulation, and pharmacological evaluation of a prolyl endopeptidase from Myxococcus xanthus: Implications for celiac sprue therapy. *Biotechnology and Bioengineering, 92*, 674–684.

Green, P. H., & Cellier, C. (2007). Celiac disease. *National England Journal of Medicine*, 1731–1743.

Guandalini, S., & Rose, R. (2012). *Evolving diagnostic criteria for celiac disease impact*. A publication of the University of Chicago Celiac Disease Center.

Gujral, N., Freeman, H. J., & Thomson, A. B. R. (2012). Celiac disease: Prevalence, diagnosis, pathogenesis, and treatment. *World Journal of Gastroenterology, 18*(42), 6036–6059.

Hall, N. J., Rubin, G., & Charnock, A. (2009). Systematic review: Adherence to a gluten-free diet in adult patients with coeliac disease. *Alimentary Pharmacology*, 315–330.

Han, A., Newell, E. W., Glanville, J., Fernandez-Becker, N., Khosla, C., Chien, Y. H., & Davis, M. M. (2013). Dietary gluten triggers concomitant activation of $CD4^+$ and $CD8^+$ $\alpha\beta$ T cells and gamma delta T cells in celiac disease. *Proceedings of the National Academy of Sciences of the United States of America*, 13073–13078.

Haush, F., Shan, L., Santiago, N. A., Gary, G. M., & Khosla, C. (2002). Intestinal digestive resistance of immunodominant gliadin peptides. *American Journal of Physiology. Gastrointestinal and Liver Physiology, 283*, 996–1003.

Hollon, J., Puppa, E. L., Greenwald, B., Goldberg, E., Guerrerio, A., & Fasano, A. (2015). Effect of gliadin on permeability of intestinal biopsy explants from celiac disease patients and patients with nonceliac gluten sensitivity. *Nutrients, 7*(3), 1565–1576.

Kale, S., Hanai, J., Chan, B., et al. (2005). Microarray analysis of in vitro pericyte differentiation reveals an angiogenic program of gene expression. *The FASEB Journal, 19*(2), 270–271.

Kapoerchan, V. V., Wiesner, M., Overhand, M., van der Marel, G. A., Koning, F., & Overkleeft, H. S. (2008). Design of azidoproline containing gluten peptides to suppress CD4+ T-cell responses associated with celiac disease. *Bioorganic and Medical Chemistry*, 2053–2062.

Karell, K., Louka, A. S., Moodie, S. J., Ascher, H., Clot, F., Greco, L., Ciclitira, P. J., Sollid, L. M., & Partanen, J. (2003). HLA types in celiac disease patients not carrying the DQA1*05-DQB1*02 (DQ2) heterodimer: Results from the European Genetics Cluster on Celiac Disease. *Human Immunology, 64*, 469–477.

Kelly, C., Bai, J., Liu, E., & Leffler, D. (2015). Celiac disease: Clinical spectrum and management. *Gastroenterology*, 1175–1186.

Khosla, C. (2017). Celiac disease: Lessons for and from chemical biology. *ACS Chemical Biology, 12*(6), 1455–1459.

Kim, C. Y., Quarsten, H., Bergseng, E., Khosla, C., & Sollid, L. M. (2004). Structural basis for HLA-DQ2-mediated presentation of gluten epitopes in celiac disease. *Proceedings of the National Academy of Science*, 4175–4179.

Lai, T. S., Liu, Y., Tucker, T., Daniel, K. R., Sane, D. C., Toone, E., Burke, J. R., Strittmatter, W. J., & Greenberg, C. S. (2008). Identification of chemical inhibitors to human tissue transglutaminase by screening existing drug libraries. *Chemistry and Biology*, 969–978.

Lai, T. S., Davies, C., & Greenberg, C. S. (2010). Human tissue transglutaminase is inhibited by pharmacologic and chemical acetylation. *Protein Science, 19*, 229–235. https://doi.org/10.1002/pro.301

Lanzini, A., Lanzarotto, F., Villanacci, V., Mora, A., Bertolazzi, S., Turini, D., Carella, G., Malagoli, A., Ferrante, G., Cesana, B. M., et al. (2009). Complete recovery of intestinal mucosa occurs very rarely in adult coeliac patients despite adherence to gluten-free diet. *Alimentary Pharmacology, 29*, 1299–1308.

Lee, A. R., Ng, D. L., Zivin, J., & Green, P. H. (2007). Economic burden of a gluten-free diet. *Journal of Human Nutrition and Dietetics*, 423–430.

Leffler, D. A., Edwards-George, J., Dennis, M., Schuppan, D., Cook, F., Franko, D. L., Blom-Hoffman, J., & Kelly, C. P. (2008). Factors that influence adherence to a gluten-free diet in adults with celiac disease. *Digestion Science*, 1573–1581.

Lequin, R. M. (2005). Enzyme immunoassay (EIA)/enzyme-linked immunosorbent assay (ELISA). *Clininal Chemistry*, 2415–2418.

Lerner, P., Jeremias, S., Neidhofer, T., & Matthias. (2017). Comparison of the reliability of 17 celiac disease associated bio-markers to reflect intestinal damage. *Journal of Clinical Cell Immunology, 8*, 1000486.

Lionetti, E., & Catassi, C. (2011). New clues in celiac disease epidemiology, pathogenesis, clinical manifestations, and treatment. *International Review in Immunology*, 219–231.

Liu, S., Cerione, R. A., & Clardy, J. (2002). Structural basis for the guanine nucleotide-binding activity of tissue transglutaminase and its regulation of transamidation activity. *Proceedings of the National Academy of Science, 99*, 2743–2747.

Lo, W., Sano, K., Lebwohl, B., Diamond, B., & Green, P. H. (2003). Changing presentation of adult celiac disease. *Digtial Science, 48*, 395–398.

Lorand, L. (2007). Crosslinks in blood: Transglutaminase and beyond. *The FASEB Journal, 21*, 1627–1632.

Lorand, L., & Graham, R. M. (2003). Transglutaminases: Crosslinking enzymes with pleiotropic functions. *Nature Reviews Molecular Cell Biology*, 140–156.

Louka, A. S., Nilsson, S., Olsson, M., Talseth, B., Lie, B. A., Ek, J., Gudjonsdottir, A. H., Ascher, H., & Sollid, L. M. (2002). HLA in coeliac disease families: A novel test of risk modification by the "other" haplotype when at least one DQA1*05-DQB1*02 haplotype is carried. *Tissue Antigens*, 147–154.

Margaritte-Jeannin, P., Babron, M. C., Bourgey, M., Louka, A. S., Clot, F., Percopo, S., Coto, I., Hugot, J. P., Ascher, H., Sollid, L. M., Greco, L., & Clerget-Darpoux, F. (2004). HLA-DQ relative risks for coeliac disease in European populations: A study of the European Genetics Cluster on Coeliac Disease. *Tissue Antigens, 63*, 562–567.

Marsh, M. N., & Crowe, P. T. (1995). Morphology of the mucosal lesion in gluten sensitivity. *Baillière's Clinical Trials in Gastroenterology, 9*(2), 273–293.

Marti, T., Molberg, O., Li, Q., Gray, G. M., Khosla, C., & Sollid, L. M. (2005). Prolyl endopeptidase-mediated destruction of T cell epitopes in whole gluten: Chemical and immunological characterization. *Journal of Pharmacology Experiments, 312*, 19–26.

Mazzarella, G., Maglio, M., Paparo, F., Nardone, G., Stefanile, R., Greco, L., van de Wal, Y., Kooy, Y., Koning, F., Auricchio, S., & Troncone, R. (2003). An immunodominant DQ8 restricted gliadin peptide activates small intestinal immune response in in vitro cultured mucosa from HLA-DQ8 positivebut not HLA-DQ8 negative coeliac patients. *Gut, 52*, 57–62.

Miura, N., Yamamoto, M., Fukutake, M., et al. (2005). Anti-celiac disease 3 induces biphasic apoptosis in murine intestinal epithelial cells: Possible involvement of the Fas/Fas ligand system in different T cell compartments. *International Journal of Immunology*, 513–522.

Molberg, McAdam, S. N., & Sollid, L. M. (2000). Role of tissue transglutaminase in celiac disease. *Journal of Pediatric Gastroenterology and Nutrition*, 232–240.

Molberg, O., Solheim Flaete, N., Jensen, T., Lundin, K. E., Arentz-Hansen, H., Anderson, O. D., Kjersti Uhlen, A., & Sollid, L. M. (2003). Intestinal T-cell responses to high-molecular-weight glutenins in celiac disease. *Gastroenterology*, 337–344.

Moss, S. F., Attia, L., Scholes, J. V., Walters, J. R., & Holt, P. R. (1996). Increased small intestine apoptosis in celiac disease. *Gut, 39*(6), 811–817.

Myrsky, E., Kaukinen, K., Syrjänen, M., Korponay-Szabó, I. R., Mäki, M., & Lindfors, K. (2008). Coeliac disease-specific autoantibodies targeted against transglutaminase 2 disturb angiogenesis. *Clinical and Experimental Immunology*, 111–119.

Naluai, A. T., Nilsson, S., Gudjonsdottir, A. H., Louka, A. S., Ascher, H., Ek, J., Hallberg, B., Samuelsson, L., Kristiansson, B., Martinsson, T., Nerman, O., Sollid, L. M., & Wahlstrom, J. (2001). Genome-wide linkage analysis of Scandinavian affected sib-pairs supports presence of susceptibility loci for celiac disease on chromosomes 5 and 11. *European Journal of Human*, 938–944.

Nasr, I., Leffler, D. A., & Ciclitira, P. J. (2012). Management of celiac disease. *Gastrointestinal Endoscopy Clinical Nutrition*, 695–704.

Niewinski, M. M. (2008). Advances in celiac disease and gluten-free diet. *Journal of the American Dietetic Association*, 661–672.

NIH Consensus Development Conference on Celiac Disease. NIH Consens State Sci Statements. Available online: http://consensus.nih.gov/2004/2004celiacdisease118.html. Accessed 21 Nov 2013.

Olsson, C., Hörnell, A., Ivarsson, A., & Sydner, Y. M. (2008). The everyday life of adolescent coeliacs: Issues of importance for compliance with the gluten-free diet. *Journal of Human Nutrition and Diet*, 359–367.

Palejwala, A. A., & Watson, A. J. M. (2000). Apoptosis and gastrointestinal disease. *Journal of Pediatrica Gastroenterology Nutrition*, 356–361.

Paterson, B. M., Lammers, K. M., Arrieta, M. C., Fasano, A., & Meddings, J. B. (2007). The safety, tolerance, pharmacokinetic and pharmacodynamic effects of single doses of AT-1001 in coeliac disease subjects: A proof of concept study. *Alimentary Pharmacology, 26*(5), 757–766.

Pinkas, D. M., Strop, P., Brunger, A. T., & Khosla, C. (2007). Transglutaminase 2 undergoes a large conformational change upon activation. *Plasma Biology, 327*.

Piper, J. L., Gray, G. M., & Khosla, C. (2004). Effect of prolyl endopeptidase on digestive-resistant gliadin peptides in vivo. *Journal of Pharmacology Experiments*, 213–219.

Ploski, R., Ek, J., Thorsby, E., & Sollid, L. M. (1993). On the HLA-DQ(# 1*0501, " 1*0201)-associated susceptibility in celiac disease: a possible gene dosage effect of DQB1*0201. *Tissue Antigens*, 173–177.

Porta, R., Gentile, V., Esposito, C., Mariniello, L., & Auricchio, S. (1990). Cereal dietary proteins with sites for cross-linking by transglutaminase. *Phytochemistry*, 2801–2804.

Pyle, G. G., Paaso, B., Anderson, B. E., Allen, D. D., Marti, T., Li, Q., Siegel, M., Khosla, C., & Gray, G. M. (2005). Effect of pretreatment of food gluten with prolyl endopeptidase on gluten-induced malabsorption in celiac sprue. *Clinical Gastroenterology, 3*, 687–694.

Rashtak, S., Snyder, M. R., Pittock, S. J., & Wu, T.-T. (2011). Serology of celiac disease in gluten-sensitive ataxia or neuropathy: Role of deamidated gliadin antibody. *Journal of Neuroimmunology, 230*, 130–134.

Rossi, F., Bellini, G., Tolone, C., Luongo, L., Mancusi, S., Papparella, A., Sturgeon, C., Fasano, A., Nobili, B., Perrone, L., Maione, S., & del Giudice, E. M. (2012). The cannabinoid receptor type 2 Q63R variant increases the risk of celiac disease: Implication for a novel molecular biomarker and future therapeutic intervention. *Pharmacological Research, 66*(1), 8.

Saturni, L., Ferretti, G., & Bacchetti, T. (2010). The gluten-free diet: Safety and nutritional quality. *Nutrients*, 16–34.

Schuppan, D., Junker, Y., & Barisani, D. (2009). Celiac disease: From pathogenesis to novel therapies. *Gastroenterology*, 1912–1933.

See, J., & Murray, J. A. (2006). Gluten-free diet: The medical and nutrition management of celiac disease. *Nutrition in Clinical Practice, 21*, 1–15.

Shan, L., Molberg, O., Parrot, I., Hausch, F., Filiz, F., Gray, G. M., Sollid, L. M., & Khosla, C. (2002). Structural basis for gluten intolerance in celiac sprue. *Science, 297*, 2275–2279.

Shewry, P. R., Tatham, A. S., & Kasarda, D. D. (1992). Cereal proteins and celiac disease. In M. N. Marsh (Ed.), *Coeliac disease* (pp. 305–342). Blackwell Scientific Publications.

Silvester, J. A., & Rashid, M. (2010). Long-term management of patients with celiac disease: Current practices of gastroenterologists in Canada. *Canadian Journal of Gastroenterology*, 499–509.

Simón, E., Larretxi, I., Churruca, I., Arrate, L., Bustamante, M. A., Virginia, N., María, D. P., Fernández-Gil, & Jonatan, M. (2017). *Nutritional and analytical approaches of gluten-free diet in celiac disease*. Springer briefs in food, health, and nutrition. ISBN 978-3-319-53342-1.

Singh, U. S., Erickson, J. W., & Cerione, R. A. (1995). Identification and biochemical characterization of an 80 kilodalton GTP-binding transglutaminase from rabbit liver nuclei. *Biochemistry, 34*, 15863–15871.

Solid, L. M. (2000). Molecular basis of celiac disease. *Annual Review of Immunology*, 53–81.

Sollid, L. M., & Lundin, K. E. A. (2009). Diagnosis and treatment of celiac disease. *Mucosal Immunology*, 3–7.

Stamnaes, J., Pinkas, D. M., Fleckenstein, B., Khosla, C., & Sollid, L. M. (2010). Redox regulation of transglutaminase 2 activity. *Journal of Bio-Chemistry*, 25402–25409.

Stevens, L., & Rashid, M. (2008). Gluten-free and regular foods: A cost comparison. *Canadian Journal of Dietetary Practice Research*, 147–150.

Tripathi, A., Lammers, K. M., Goldblum, S., Shea-Donohue, T., Netzel-Arnett, S., Buzza, M. S., Antalis, T. M., Vogel, S. N., Zhao, A., Yang, S., Arrietta, M. C., Meddings, J. B., & Fasano, A. (2009). Identification of human zonulin, a physiological modulator of tight junctions, as prehaptoglobin-2. *Proceedings of the National Academy of Science United States of America*, 16799–16804.

Wood Heickman, L. K., DeBoer, M. D., & Fasano, A. (2020). Zonulin as a potential putative biomarker of risk for shared type 1 diabetes and celiac disease autoimmunity. *Diabetes Metabolism Research*, 363–309.

Xia, J., Bergseng, E., Fleckenstein, B., Siegel, M., Kim, C. Y., Khosla, C., & Sollid, L. M. (2007). Cyclic and dimeric gluten peptide analogues inhibiting DQ2-mediated antigen presentation in celiac disease. *Medical Chemistry*, 6565–6573.

Yokoyama, S., Watanabe, N., Sato, N., Perera, P. Y., Filkoski, L., Tanaka, T., Miyasaka, M., Waldmann, T. A., Hiroi, T., & Perera, L. P. (2009). Antibody-mediated blockade of IL-15 reverses the autoimmune intestinal damage in transgenic mice that overexpress IL-15 in enterocytes. *Proceedings of the National Academy of Sciences of the United States of America*, 15849–15854.

Zarkadas, M., Cranney, A., Case, S., Molloy, M., Switzer, C., Graham, I. D., Butzner, J. D., Rashid, M., Warren, R. E., & Burrows, V. (2006). The impact of a gluten-free diet on adults with celiac disease: Results of a national survey. *Journal of Human Nutrition and Dietetics*, 41–49.

# Chapter 2
# Nutritional Aspects and Health Implications of Gluten-Free Products

Check for updates

**Surabhi Pandey**

**Abstract** Celiac disease (CD) is an autoimmune enteropathy arising from the peculiar immune response to gluten-derived peptide amongst the susceptible population. It is evidenced by the chronic inflammation of the mucosal surface and atrophy of the intestinal villi, resulting in abnormal absorption of nutrient. The pathogenesis of CD involves the molecular interaction between gluten peptides, intestinal epithelium, and T-lymphocyte cells, the activity of the latter being enhanced by transglutaminase located at the epithelial brush border. Gluten is a protein that attributes to the viscoelastic properties of dough and enhances the gas retention and structure of the baked products. It constitutes a composite of cereal storage proteins including prolamins and glutenins. The toxic prolamins in wheat, barley, and rye consist of gliadin, hordein, and secalin, respectively. These prolamins have a high amount of proline and glutamine that resists degrading in the gastrointestinal environment, which consequently agglomerate as large peptide fragments. The toxic protein fragments induce mucosal damage and activate the T-lymphocyte cells which in turn produces high levels of pro-inflammatory cytokines causing clonal expansion, thus depicting the hallmark of CD. The aim of the chapter is to discuss the idea of nutritional Aspects of Gluten-Free Products.

**Keywords** Enteropathy · Prolamines · Celiac diseases · Gluten-free

## 2.1 Introduction

As per the definition of Codex Alimentarius standards, the term gluten-free (GF) refers to foods comprising gluten under the permissible value of 20 ppm (Standard, 2007). The key aspect for safe consumption of the GF diet is the absence of gluten in natural or processed foods. Many of the gluten-containing cereals (wheat, barley, and rye) and their hybrids (spelt, triticale, semolina, malt, etc.) are restricted for the

S. Pandey (✉)
Bioresource Engineering Department, McGill University, Ste-Anne-de-Bellevue, Canada
e-mail: surabhi.pandey@mail.mcgill.ca



N. Singh Deora et al. (eds.), *Challenges and Potential Solutions in Gluten Free Product Development*, Food Engineering Series,
https://doi.org/10.1007/978-3-030-88697-4_2

celiac population. Up to now, the only therapy to combat CD, is strict adherence to a GF diet. GF diet mainly consists of the consumption of GF cereals, pseudocereals, fruits, vegetables, pulses, meats, and specially produced GF products in which gluten is replaced by GF flours. Commonly used substitutes for the gluten-containing cereals include rice and corn, followed by sorghum and oats. The use of oats as a gluten substitute is questionable due to the presence of avenin (storage protein), however, some studies have confirmed that oats can be well digested by most of the celiac population, and improves the palatability and nutritional value (Lee, 2009). Other than that, less commonly grown cereals, also called minor cereals (such as teff, and millet) and pseudo-cereals which are small grain-like seeds (buckwheat, quinoa, and amaranth) represent another possibility (Alvarez-Jubete et al., 2010).

Gluten network engulfs the starch granules and prevents its easy access to the amylase, thus slowing down the starch hydrolysis rate in the small intestine. The removal of gluten from the products increases the postprandial blood glucose level in the body leading to obesity and metabolic disorders (Scazzina, 2015). Consequently, a major fraction of celiac patients has shown nutrient deficiencies including that of calorie/protein, dietary fiber, minerals, and vitamins (Bardella, 2000; Thompson, 2000; Barton et al., 2007). The incidence of iron and vitamin B12 deficiency in celiac patients varies from 12% to 69%, and 8% to 41%, respectively (Tikkakoski et al., 2007; Dahele & Ghosh, 2001). Several reports showed that GF diets are hyperproteic and hyperlipidic and do not provide an adequate amount of carbohydrates, calcium, iron, and fiber, resulting in overweight conditions amongst CD patient. Many of the GF products contain trans fatty acids and dietary lipids that trigger metabolic imbalance and risk of coronary heart disease (Lissner & Heitmann, 1995). At the same time, vitamin D deficiency may develop due to the avoidance of lactose from milk and dairy products, as the CD patients become secondary lactose intolerance due to the reduced lactase production by the damaged intestinal villi. However, the severity of the CD related nutrient deficiencies depends on the several factors such as degree of malabsorption, the severity of intestinal damage, and length of undiagnosed period for an active CD patient (Ojetti, 2005).

While several studies have suggested that adhering to the GF diet can resolve the nutrient deficiency (Hallert, 2002; Annibale, 2001), some authors have argued that following a strict GF diet can mitigate nutritional deficiencies. Henceforth, this chapter deals with the different aspects of nutritional characteristics of GF ingredients and the GF products keeping health perspective into consideration.

## 2.2 Nutritive Profile and Bioactive of Gluten-Free Ingredients

### *2.2.1 Major Gluten-Free Cereals*

Currently, the most widely used cereal flours for making GF products include rice (*Oryza sativa*) and maize or corn (*Zea mays*) owing to their hypoallergenic character, bland flavor, and easy availability (Kadan, 2001). The nutritional composition and amino acid profile of gluten-containing, and GF grains suggest that that rice and corn have lesser protein content and total dietary fiber, while carbohydrate amount was considerably high. The rice proteins are insoluble due to their hydrophobic nature, due to which rice flour demonstrates inferior viscoelastic properties during the baking process. The carbon dioxide produced during fermentation escapes due to poor protein-starch network, as a result of which the product formed is compact and brittle with poor sensory qualities (Marco & Rosell, 2008). Maize is a high energy crop, but at the same time has low levels of essential amino acids such as tryptophan and lysine and lacks important minerals and vitamins (Foschia, 2016).

Sorghum (*Sorghum vulgare*), on the other hand, has higher protein content than rice and comparable to that of maize. At the same time, it consists of a comparatively lower carbohydrate than the two former cereals. The release of sugars from the starch matrix is relatively slower to other cereals, which makes it an ingredient of interest to diabetic and celiac people. Due to the poor starch and protein digestibility, a pretreatment process is usually carried out to release the components from the compact complexes (Correia, 2010). These processes include fermentation, malting, and enzymatic treatment. There has been increased attention in utilizing sorghum as a sole GF ingredient or in combination with other non-gluten flours. The major protein fraction present in sorghum consists of globulins and prolamins that are generally surrounded by starch granules (Marti & Pagani, 2013).

The use of oats (*Avena sativa*) as a GF ingredient is still controversial as it is believed that avenins (storage proteins in oats) can trigger up the immunogenic response. At the same time, others suggest that the immunogenicity depends on the cultivar consumed. Nevertheless, it has numerous health benefits such as high unsaturated fatty acids and dietary fibers mainly β-glucans (Lasa, 2017).

## 2.3 Minor Gluten-Free Cereals

### *2.3.1 Millet*

Millet is characterized as small-seeded coarse cereal and is the most extensively studied cereal after rice, corn, and sorghum. They are categorized as the underutilized food in North America and Europe; however, their drought-resistant nature and low agricultural inputs make them a suitable crop for India, Africa, and China. The different varieties of millet grown today are Finger millet (*Eleusine coracana*), Foxtail millet (*Setaria italica*), Pearl millet (*Pennisetum glaucum*), Kodo millet (*Paspalum setaceum*), and Proso millet (*Panicum miliaceum*). Adding to these advantages, they have a relatively lower glycemic index and exceptional nutritional benefits (Saleh, 2013). Apart from being a gluten-free alternative, it has been helpful for the management of type II diabetes owing to their hypoglycemic properties (Annor, 2017). They suggested that the presence of lipids, proteins, α-amylase inhibitors, starch type, and phenolic compounds are the contributing factors to the hypoglycemic activity of millets. Most of the millet proteins contain a range of essential amino acids, with a relatively high quantity of methionine (Singh et al., 2012).

Pearl millet consists of around 69.1% carbohydrate, 11.4% crude protein, 4.8% crude fat, 2% crude fiber, and 2.2% ash. Finger millet has many health benefits, a few of which have been attributed to the presence of phenolics. The nutritional value of finger millet is marked as 71.52% carbohydrate, 7.44% crude protein, 3.6% crude fiber, and 3.38 mg/g of calcium. Foxtail millet, on the other hand, has a significant amount of lysine which makes them an additional protein source for most of the cereals (Ragaee et al., 2006). Kodo millet and little millet have fats mostly containing polyunsaturated fatty acids; are found to have 38% dietary fiber which is highest amongst other cereals (Hegde et al., 2005). Compared to the protein profile present in wheat, proso millet contains more essential amino acids such as methionine, leucine, and isoleucine (Kalinova & Moudry, 2006). The highest amount of protein and crude fiber is found in Proso millet, while finger millet possesses the highest amount of calcium suitable for fighting anemi (Chethan & Malleshi, 2007; Sripriya et al., 1997). Along with specific mineral and proteins, millets also contain dietary fibers such as resistant starch that are needed for the synthesis of short-chain fatty acids (butyrate) effective in preventing colon cancer. The in-vitro study on soluble polysaccharides (arabinose and xylose) have proved their prebiotic activity and wound dressing property (Mathanghi, 2012).

Speaking of micronutrients, millets are an excellent source of β-carotene and B-vitamins especially folic acid, niacin, and riboflavin. The amount of thiamine and niacin present in millet is comparable to that of rice and wheat. Besides, millet's bioactive constituents complement those present in fruits and vegetables. These include gallic acid, catechol, cinnamic acid, benzoic acid, ascorbic acid, ferulic acid, *p*-coumaric acid, vanillic acid, sinapic acid, chlorogenic acid, proto-catechuic acid, kaempferol, caffeic acid gentisic acid, salicylic acid, and syringic acid, however, the concentration of these bioactive compounds vary according to the cultivar and environmental conditions (Kumar, 2018).

### *2.3.2 Teff*

**Teff** (*Eragrostis tef*) belongs to the Poaceae family, and is a minor GF cereal produced in Ethiopia and Eritrea (*Eragrostis teff*), and exhibit an excellent protein profile (Bultosa & Teff, 2016). Teff seeds are mostly distinguished based on color (white, red or brown, and mixed) for marketing purposes. The color of the hulled grains ranges from pale white to dark brown (Belay, 2009). The starch content in teff (73%) is relatively higher than most of the cereals which makes it a suitable alternative to wheat (Tatham, 1996). The protein profile of teff is similar to wheat and higher than rice, maize, sorghum, and millet.

The major storage proteins present in teff are glutelins (44.55%) and albumins (36.6%) and the rest constitute prolamins (11.8%) and globulins (6.7%) (Tatham,

1996). Since the teff flour consists of albumin, glutelin, and globulin as the major protein fractions, it is easily digestible relative to other gluten-free cereal such as sorghum and maize (Gebremariam et al., 2014).

The mean value of crude fat present in teff is 0.0238%, out of which 72.46% is unsaturated fatty acids, mainly consisting of oleic acid (32.41%) and linoleic acid (23.83%) (El-Alfy et al., 2012). The crude fiber content of teff is analogous to that of millets and higher than rice, wheat, maize, and sorghum. The reason for such high fiber content is its exceptionally small grain size due to which it is always milled as whole-grain flour.

Moreover, calcium which helps in bone building and prevention of colon cancer is present in high amounts in teff. To prevent the issues related to the low calcium intake, Roosjen suggested that flour should contain at least 150 mg/100 g of calcium (Roosjen, 2005). Except for teff and finger millet, most of the major cereals such as rice, wheat, maize, and sorghum fail to fulfill this requirement. Apart from that, teff has been found reasonable for the sparse occurrence of anemia in Ethiopia. Other than niacin, riboflavin, and thiamine; teff also contains vitamin B6 (0.482 mg/100 g), vitamin K (phylloquinone), vitamin A (9 IU), and α-tocopherol (0.08 mg/100 g) (wet basis) (Zhu, 2018).

Similar to millets, teff exhibits health-promoting effects due to the presence of a substantial amount of phenolics. The major phenolic compound present in teff includes ferulic acid (285.9 μg/g), with considerable amounts of cinnamic (46 μg/g), vanillic (54.8 μg/g), coumaric (36.9 μg/g), protocatechuic (25.5 μg/g), gentisic (15 μg/g), syringic (14.9 μg/g) acids (McDonough et al., 2000). These phenolics are responsible for the antioxidant activity, which helps in the prevention of cardiovascular diseases and cancer (Awika & Rooney, 2004).

### *2.3.3 Pseudo-cereals*

In contrast to monocotyledonous cereals, Buckwheat (*Fagopyrum esculentum*), Amaranth (*Amaranthus* spp.), and Quinoa (*Chenopodium quinoa*) are dicotyledonous seeds under the family of *Poaceae*. The common term assigned to them is pseudo-cereals as they have similar structural and nutritional properties to that of the true cereals. These pseudo cereals present good opportunities for the production of GF products as they not only lack toxin prolamins; but are characterized by high macronutrients and micronutrients including the essential amino acids. Amaranth is a lens-shaped seed with a diameter varying from 1 to 1.5 mm and weighs around 0.6 to 1.3 mg (Bressani, 1994). The common amaranth species under cultivation are *Amaranthus hypochondriacus, Amaranthus caudatus, and Amaranthus cruentus.* Compared to Amaranth seeds, quinoa seeds are much bigger with a diameter of 1–2.5 mm (Taylor & Parker, 2002). The popular buckwheat varieties are common buckwheat (*Fagopyrum esculentum* Moench) and tartary buckwheat (*Fagopyrum tataricum*) (Oomah & Mazza, 1996). *The nutritional compositions of these pseudo-cereals from where it can be derived that amaranth and quinoa have the highest protein amongst all the cereals. The main advantage of using pseudo-cereals as the gluten-free ingredient is in the fact that they contain globulins and albumins as the main protein fractions, with negligible prolamins proteins, the latter being toxic to the celiac patients(Drzewiecki, 2003). The amino acid present in globulins and albumins fractions contains a lower amount of* proline and glutamic acid than prolamines (Gorinstein, 2002). These pseudo-cereals are a decent source of dietary fiber; therefore, infusion of these seeds with other gluten-free ingredients can help to alleviate the dietary fiber deficiency amongst the concerned celiac population (Alvarez et al., 2009)

Another important nutritional aspect of the pseudo-cereals is the presence of a high percentage of unsaturated fat such as linoleic acid (50% of the total fatty acid for amaranth and quinoa, and 35% for buckwheat), oleic acid (25% for amaranth and quinoa, and 35% for buckwheat), and palmitic acid (Bruni, 2001; Bonafaccia et al., 2003). The high α-linolenic content (3.8–8.3%) in quinoa seeds is responsible for the reduction of biological markers linked to common degenerative diseases namely, cancer, osteoporosis, cardiovascular, and autoimmune diseases (Ruales & Nair, 1993)

Furthermore, buckwheat consists of fagopyritols, a soluble carbohydrate which is a source of D-chiro-inositol, a compound that effectively controls the type II diabetes through glycemic index management. The range of fagopyritols present in buckwheat ranges from 269.4 to 464.7 mg/100 g (Steadman, 2000). The main phenolic compounds present in amaranth seeds are ferulic acid, caffeic acid, and p-hydroxybenzoic acid (Klimczak et al., 2002). Flavonoids such as glucosides of flavonol kaempferol and quercetin are abundantly found in quinoa seeds (Dini et al., 2004). For buckwheat, the main source of phenolics is glucosides of flavonol quercetin, accompanied by glycosides of the flavones luteolin and apigenin (Dietrych-Szostak & Oleszek, 1999).

## 2.4 Legumes

Besides pseudocereals, several attempts have been made to utilize legumes (pulses, in particular) to improve the protein profile and functional characteristics of GF products. Typically, pulses are categorized into 11 main classes: dry beans (*Phaseolus* spp.; *Vigna* spp.), dry peas (*Pisum* spp.), dry broad beans (*Vicia faba*), dry cowpeas or black-eyed peas (*Vigna unguiculata* (L.) Walp.), chickpeas or bengal grams (*Cicer aretinum* L.), pigeon peas (*Cajanus cajan* L. Mill sp.), bambara groundnuts or earth peas (*Vigna subterranean* L.), several varieties of lentils (*Lens culinaris* Medik.), vetch or common vetch (*Vicia sativa* ), and lupins (such as *Lupinus albus* L. and *Lupinus mutabilis* Sweet), as well as minor pulses, including jack beans (*Canavalia ensiformis*), lablab or hyacinth beans (*Lablab purpureus* ), winged beans (*Psophocarpuster agonolobus* ), sword beans (*Canavalia gladiata* ), yam beans (*Pachyrrizus erosus* ), and velvet beans (*Mucuna pruriens* var. *utilis* ) (Melini, 2017). These pulses contain a good amount of protein, complex carbohydrates, dietary fibers, and micronutrients. At the same time, they contain a high quantity of polyphenols demonstrating excellent antioxidant property, and other secondary metabolites including isoflavones, bioactive carbohydrates, polysterols, saponins, and alkaloids (Roy et al., 2010) Besides, pulses based GF products have low glycemic index.

The main protein proportion in pulses accounts for about 17–35% of the cotyledon weight (dry basis) (McCrory, 2010). Pulse proteins are categorized into four classes based on their solubility in several solvents i.e. (a) albumins (water-soluble), (b) globulins (soluble in a salt solution), (c) prolamins (soluble in an alcoholic solution of 70% concentration), and d) glutelins (dissolves in alkaline solution), of which globulins and albumins constitute the major proportion (60–80%) of protein fraction (Tiwari & Singh, 2012). Compared to cereal proteins, pulses possess higher levels of leucine, lysine, arginine, aspartic, and glutamic acid, however contain lower amounts of sulfur-containing amino acids such as cysteine, and methionine. Based on the protein concentration, pulses flours are available as pulse flour (<65%, db), protein concentrates (>65%, db), and protein isolates (>90%, db) (Schoenlenchner, 2010).

The carbohydrates in pulses are composed of starch, soluble sugars, and dietary fibers, which overall account for 55–65% of the whole pulse dry weigh (Boye et al., 2010). The starch in total constitutes of 22–45% of total carbohydrates, and generally contains up to 20–30% amylose and 70–80% amylopectin (Maaran, 2014). The resistant starch present in pulses attributes to the slow glucose release thus, controlling the glycemic and postprandial responses (Berrios et al., 2010). Along with that, they also represent a source of dietary fiber (soluble and insoluble, both), for instance, the insoluble fiber in chickpeas, lentils, and peas are 75%, 87%, and 89% (db) (Maaran, 2014).

Furthermore, they are a rich source of niacin, thiamine, riboflavin, pyridoxamine, pyridoxal, and pyridoxin. High amounts of folates are also present in pulses which are generally deficient in humans due to complex bonds with other biomolecules. As a relevant example, beans contain 400–600 μg/100 g of folates which covers 95% of the daily requirement. Chickpeas are also rich in folates and contain a higher amount than peas (150 vs 102 μg/100 g) (Campos et al., 2010). They also contain a good amount of minerals such as iron, zinc (highest in beans and lentils), potassium, and magnesium (highest in cowpea) (Oomah, 2011). Amongst secondary metabolites such as tannins, phenolic acids, and flavonoids, kidney beans and black grams contains the highest phenolic content. Chickpeas also contain a wide variety of bioactives including glucosides of flavone, flavonoids, and oligomeric cum polymeric proanthocyanidins.

## 2.5 Nutritional Interventions in GF Products

The production of baked products from gluten-free ingredients has several associated challenges. Firstly, the viscoelastic property of gluten is hard to mimic in order to develop a palatable GF baked product. During baking, gluten plays an important role in holding the $CO_2$ released during proofing, thus giving the product an excellent structure and acceptable volume (Drabinska et al., 2016). The absence of gluten impairs the dough structure, giving a liquid consistency batter and defected baked products (Gallagher et al., 2014). For instance, the removal of gluten prevents starch swelling during cooking of pasta, prevents the biological leavening, and hampers the texture of bread, whereas for biscuits it has an impact on the elasticity and cohesiveness of the dough (Di Cairano, 2018). Another challenge is to maintain an appropriate nutritional profile for these kinds of products. Frequently, the gluten-free products are characterized by a high amount of saturated fatty acids and sugars, and low levels of nutrients such as dietary fibers, calcium, iron, zinc, magnesium, vitamin B12, and folate. Moreover, GF products often depict a high glycemic index due to the presence of starchy ingredients. The high GI leads to several metabolic disorders such as obesity and diabetes (Jnawali et al., 2016; Naqash, 2017; Vici, 2016).

The most common cereals used as ingredients for GF products are rice and corn which have certain demerits associated such as poor viscoelastic properties of rice which hinders gas retention during baking, and inferior textural properties of GF product when developed using cornflour. On the other side, the protein and starch present in sorghum are not easily digestible. The high gelatinization temperature of sorghum flour results in poor quality bread with cracks and large holes in the crumb (Carcea, 2020). Therefore, substances such as hydrocolloids (guar and xanthan gums, alginate, carrageenan, hydroxypropyl methylcellulose, carboxymethyl cellulose), emulsifier, isolated proteins (from egg, legumes, or dairy products), or enzymes (cyclodextrin glycosyltransferases, transglutaminase, proteases, glucose oxidase, and laccase) have been added during the preparation of GF products to mimic the property of gluten (Matos & Rosell, 2015) Other interventions, sourdough technology, and high hydrostatic pressure being some of them, have been applied to improve the organoleptic qualities and nutritional properties, consequently (Capriles et al., 2016).

## 2.6 Baked Products

### 2.6.1 *Biscuits and Cookies*

The most commonly employed GF alternatives for biscuit preparation are maize starch and rice flour. However, these flours give high energy intake but lack many essential amino acids (lysine and tryptophan) and vitamins. Thus, to surpass the low nutrient challenge, different combinations of high-value ingredients have been used to enhance the nutritional properties, bioactive content, and glycemic index of GF biscuits. Rybicka and Gliszczynska-Swiglo (2017) suggested that biscuits prepared using buckwheat chickpea, oats, millet, amaranth, teff, and quinoa had higher mineral content than those prepared using potato, maize, rice, and GF wheat starch (Rybicka et al., 2019). The technological processes such as malting, fermentation, and germination can impact the overall quality of the biscuit. A relevant example of this was presented by Omoba et al. (2015) where they used sourdough technology to improve the nutritional profile of sorghum, pearl millet, and soy-based biscuits. They observed a moderate increase in phenolic content, consequently raising the antioxidant activity than the control and a reduction in the concentration of antinutritional phytate. At the same time, the addition of antioxidants, ash, and fibers rich ingredients can have a negative impact on the optical and sensory quality of GF biscuits, therefore, giving a darker colored product with a bitter aftertaste (Omoba et al., 2015).

In addition, the use of low fiber starch such as refined flour induces a high glycemic index in the developed product. It has been observed that the utilization of wholegrain flour from legumes or pseudo-cereals, and the high moisture thermal treatments (annealing) can reduce the glycemic index of the product (Rocchetti et al., 2018; Giubrerti & Gallo, 2018). For example, the biscuits prepared using tartary buckwheat presented a lower glycemic index (62.8) than that prepared using rice flour (110.2). Furthermore, the use of malted tartary buckwheat flour can boost up the antioxidant activity, with a further reduction in the glycemic index value to 57.6 (Molinari, 2018). Likewise, teff flour can also lead to the reduction of GI when compared with other conventional GF alternatives. The incorporation of soluble fibers, such as arabinoxylans, guar gum, high molecular weight β-glucan, or psyllium can significantly lower the glycemic index by delaying the gastric functioning (Scazzina et al., 2013). The carbohydrate present in a food is indicated by the glycemic load, which can be reduced by the increased concentration of non-digestible carbohydrates like resistant starch, as well as protein content.

Adebiyi (2017) demonstrated that fermentation and malting improved the nutritional characteristics of the pearl millet biscuits; which was pointed out by the increase in the amino acids, total phenolic and mineral content (Abediyi, 2017). Another study showed that using germinated flour blends of foxtail, kodo, and barnyard millets contained higher protein content, total antioxidant activity, and phenolic content than native blends (Sharma et al., 2016). Teff, on the other side, is characterized by high protein content, but it lacks gluten which impairs the biscuit quality. Oats can improve the nutritional properties of the conventional gluten-free biscuits by partial or full replacement of the GF flour. Incorporation of oats bran in oats biscuits increased the nutritive values and dietary fiber content (Duta & Culetu, 2015).

The most studied pseudo-cereal for GF formulations is buckwheat flour which has a peculiar phytochemicals present called rutin. Some studies have shown that the incorporation of buckwheat flour with rice or wheat flour has raised the DPPH radical scavenging activity, mineral content, total phenolic content, and rutin levels than those of control biscuits (Sedej, 2011).The addition of quinoa flakes (25%) and flour (30%) into base material (corn starch) lead to the increment in the dietary fiber content. The total fatty acids present in quinoa-based cookies were composed of polyunsaturated fatty acids (60.53%), monosaturated fatty acids (23.41%), and saturated fatty acids (17.45%). The essential amino acids identified were isoleucine, methionine, threonine, phenylalanine, and valine (Brito, 2015). Chauhan et al. (2015) compared the GF cookies prepared using native and germinated amaranth flour, wherein germinated GF biscuits exhibited higher antioxidant activity and total dietary fiber than raw amaranth flour biscuits (Chauhan et al., 2015).

At the same time, legume flour has been applied in the GF products to increase their nutritional quality. Sparvoli (2016) used a low anti-nutritional variety of common beans in combination with maize flour to make low glycemic index nutritive GF biscuits. The actual glycemic index value of the prepared biscuit was higher than the predicted one due to the presence of α-amylase inhibitors (Sparvoli, 2016).

### *2.6.2 Bread*

A large section of the human population depends on bread as a staple meal which serves as a vital source of protein. However, the absence of gluten rendered low protein bread with poor sensory quality. The average protein value of GF bread is 4.4 g/100 g which is significantly less than that of conventional gluten-containing bread (10 g/100 g) (Do et al., 2014). Using protein isolates and protein-enriched flour are suitable options to augment the protein content of GF bread; however, they have an antagonistic impact on the texture and sensory characteristics of bread. Apart from gluten, the dough's viscoelasticity changes with the amylose content of the starch (Kaur, 2015). Rice flour is a common alternative for the replacement of conventional bread; however, they provide comparatively lower protein (6.14–7.30 g/100 g) and nutrition (Molina-Rosell & Matos, 2015). Leguminous flours (soybean, peas, lentils, chickpea, and beans), on the other hand, provide better protein content and nutritional profile than rice or maize flour. The addition of pea, chickpea, lentil, or pea flour with rice flour in a 1:1 ratio increased the protein content of the cake from 6.2 to 8.7 g/100 g. The addition of soy flour to starch of different origin (such as cassava, corn, and rice) improved the bread quality, sensory quality, as well as the nutritional characteristics of GF bread (Taghdir, 2017). However, the negative aspect of leguminous flour resides in their extremely high-fat content and presence of anti-nutritional compounds (Molina-Rosell & Matos, 2015).

Pseudo-cereals have similar protein profiles to that of glutinous flours, as a result of which, the bread developed using amaranth flour has better protein level and health benefits. According to Kaur (2015) the bread produced using buckwheat flour showed an optimal balance of amino acids and phenolics, however, the sensory quality was rated lower than wheat flour. The blend of rice and buckwheat showed the potential to increase the textural properties of bread, without the addition of hydrocolloids (Badiu et al., 2014). In addition, bread prepared using soy/egg/corn protein isolates (Foschia, 2016; Crockett et al., 2011) egg white solids (Nunes et al., 2009), whey protein isolates/concentrates were studied for their textural, sensory properties, and consumer acceptance.

Besides, the addition of dietary fiber to refined flour or starch lowers the glycemic index of the baked products due to improved water-binding ability. Soluble dietary fibers such as psyllium and guar gum are known to reduce the glycemic index of GF bread by 40 and 41%, respectively (Scazzina, 2015). Likewise, inulin reduced the glycemic index of GF bread from 71 to 48 (Segura & Rosell, 2011). Although these additives are advantageous for the reduction of GI, they concomitantly result in hard bread crumb with poor sensory quality. Another study used the blend of plantain-chickpea-maize to produce GF bread (Flores et al., 2015).

Apart from the ingredients used, the glycemic index of the baked product also depends on pre-processing technologies such as enzymatic treatment, germination, and sourdough fermentation. Enzymatic treatment reduces the susceptibility of starch to hydrolysis in presence of α-amylase (Dura & Rosell, 2016). It should be noted that the effect of sourdough technology varies with the type of substrate. For instance, the application of sourdough technology on sorghum and teff showed the expected decrease in GI, however using the same technique for quinoa and buckwheat resulted in a high glycemic index for the baked product (Wolter, 2014). Scazzina (2015) pointed out that white sourdough bread had higher fiber and protein, but a lower glycemic index (52.1) than normal white bread having a glycemic index of 61.2. Mixed sourdough dough (6% millet, 6% buckwheat, and 21% corn flour) was formulated and made into a multigrain puffed cake with a moderate glycemic index value of 66.7.

As far as the micronutrient availability of GF products is concerned, studies have evidenced that most of these products are crafted using highly refined substrates, which in result supplies very little nutrients (Wolter, 2014; Suliburska, 2013; Stantiall & Serventi, 2018). Several studies have suggested that the addition of buckwheat, amaranth, pearl millets, and flaxseeds have been found to increase the mineral content of the bread. The addition of buckwheat flour (10–40%) in GF bread composed of corn and potato starch enriched the micronutrient profile of the bread. The Fe content increased from 42.7 to 54.3 g/100 g and Zn content improved from 5.7 to 13 g/100 g; however, the major increase of 4 and 9 times was observed for Cu and Mn, respectively) (Rozylo, 2015). Although the nutritional quality of blends consisting of teff or pseudo-cereals (amaranth and quinoa) is much better than wheat flour; their baking properties and sensory characteristic were inferior to that of the conventional flour. As an example, it was reported that macronutrients such as fat and protein, as well as the mineral content such as Ca, K, Mg, Fe, Mn, and Zn were higher for teff, amaranth, and quinoa bread compared to what was produced using wheat (Rybicka et al., 2019)

### 2.6.3 Extrudates

Out of all the gluten-free products, pasta is the most popular product amongst the gluten-intolerant people (Gimenez, 2016). One of the main nutritional benefit of pasta is the low GI which controls the body weight, plasma lipid, and blood glucos. Gluten-free grains (replacing partially or totally) such as rice and corn have been used to target the specific group of the celiac population. Even though these conventional alternative grains are composed of many macro and micronutrients, they are not considered sufficient to fulfill all the requirements for essential micronutrients. Different types of GF flours (sorghum, rice, corn, and potato starch) have been used for spaghetti preparation, wherein the best results were obtained with a 40:20:40 ratio of sorghum, rice, and potato flour (Giacco et al., 2016; Demirkesen, 2010; Ferreria, 2016).

Different sources of proteins (egg white, whey, bovine plasma, cowpea, and lupine) have been incorporated in the GF products to improve their overall nutritional quality and texture (Kumar, 2019; Chapbell et al., 2016; Furlan et al., 2015). High protein legume (faba, lentil, and black gram) flours were used to develop the low glycemic index and highly nutritious pasta, with a reduction in the antinutritional compounds such trypsin, α-galactosides, and phytic acids during extrusion processing. Compared to the commercial cereal pasta, the legume pasta has 2.9–3.5 fold higher protein content, and 1.4–1.6 fold lower starch content. Moreover, legume flour has better lysine content, while cereal proteins contain more sulfur-containing amino acids. The anti-nutritional compounds such as trypsin inhibitors, α-galactosides, and phytic acids experience a decrease during cooking). In addition, the pasta prepared using soy flour in a combination of defatted almond flour demonstrated 3–5 times higher protein content (33.3–42.1%, db) in comparison to control pasta (2.8–13%, db) (Laleg, 2016; Martinez, 2017).

Incorporating biotechnological preprocessing such as fermentation and sprouting has been found to enhance free amino acids, minerals, and bioactives. Fermentation of black gram enriched rice evidenced the enhanced nutritional and functional properties with further improvement in total phenolic content and antioxidant activity after the extrusion. Another study focused on improving the nutritional characteristics of rice-based pasta by enrichment with fermented or sprouted sorghum flour. However, the results indicated a drop in protein and starch content after fermentation and sprouting. The limited starch breakdown in fermented sorghum enriched rice pasta does not lead to foul color or textural changes and appears to have a beneficial impact on the cooking properties. The decrease in protein content was attributed to the proteolysis of non-aggregated kafirins, thus, conserving the proteins necessary to form a stable network in the final product. During sprouting, starch cleavage by amylolytic activity alongside the peculiar protein breakdown rendered severe impairment in the cooking and nutritional properties of the enriched product (Rani, 2018; Marengo, 2015).

Satisfactory results have been obtained by blending pseudo-cereals (amaranth, buckwheat, and quinoa) with corn, soy, oats, and cassava to produce GF spaghetti (Mastromatteo, 2011; Chillo, 2009; Caperuto et al., 2001; Fiorda, 2013;

Gungormulsler et al., 2007). Milling treatment was observed to affect the nutritional profile of the quinoa used for the preparation of oat-quinoa spaghetti. The rise in amino acids including histidine, leucine, isoleucine, valine, methionine was observed, while the overall protein and lipid contents were reduced by the factor of 3.5 and 5.4, respectively. Nevertheless, the addition of quinoa guaranteed the improvement of the amino acid profile for corn protein. Moreover, the teff based pasta was reported to demonstrate similar textural properties to wheat flour; however, the overall sensory quality was inferior to the latter. Teff pasta had higher mineral, fiber content, as well as the low glycemic index (47) than the wheat-based pasta (65) (Hager, 2013).

On the other hand, using green banana flour-based pasta had inferior nutritional properties compared to standard pasta. However, the GF pasta prepared using banana-cassava composite flour has a more resistant starch content compare to semolina based pasta. Egg white and soy protein were incorporated in banana-cassava based pasta, wherein soy protein gave better protein content than egg protein powder. Furthermore, a study compared the rice, legume, and pseudo-cereal based pasta, which was shown to have satisfactory phenolics components. It was found that cooking via boiling reduced the bound to free ratio of phenolic compounds for all the GF pasta (Zandonadi, 2012; Rachman, 2020; Rocchetti, 2017).

## 2.7 Conclusions

In a nutshell, the demand for GF products has been increasing immensely which are majorly prepared using refined flour, however, at the same time their nutritional profile remains a major challenge. During milling, the minerals present in the bran and germs are lost which leads to poor nutritional quality in the end product. It has been observed that inferior micronutrient profile (minerals, vitamins, and bioactive content) can be improved using amaranth, buckwheat, and quinoa. Using a combination of different flours and incorporation of sources of protein and dietary fibres has shown to improve the micronutrients, amino acid profile, and bioactive compounds in the GF products. Different pre-treatments are cooking process have also been seen to impact the changes in the macronutrients and micronutrients of the GF products.

## References

Adebiyi, J. A. (2017). Comparison of nutritional quality and sensory acceptability of biscuits obtained from native, fermented, and malted pearl millet (*Pennisetum glaucum*) flour. *Food Chemistry*, 210–217.

Alvarez-Jubete, L., Arendt, E., & Gallagher, E. (2009). Nutritive value and chemical composition of pseudocereals as gluten-free ingredients. *International Journal of Food Sciences and Nutrition*, 240–257.

Alvarez-Jubete, L., Arendt, E. K., & Gallagher, E. (2010). Nutritive value of pseudocereals and their increasing use as functional gluten-free ingredients. *Trends in Food Science and Technology*, 106–113.

Annibale, B. (2001). Efficacy of gluten-free diet alone on recovery from iron deficiency anemia in adult celiac patients. *The American Journal of Gastroenterology, 96*(1), 132–137.

Annor, G. A. (2017). Why do millets have slower starch and protein digestibility than other cereals? *Trends in Food Science and Technology*, 73–83.

Awika, J. M., & Rooney, L. W. (2004). Sorghum phytochemicals and their potential impact on human health. *Phytochemistry, 65*(9), 1199–1221.

Badiu, E., Aprodu, I., & Banu, I. (2014). Trends in the development of gluten-free bakery products. The Annals of the University Dunarea de Jos of Galati. *Fascicle VI-Food Technology*, 21–36.

Bardella. (2000). Body composition and dietary intakes in adult celiac disease patients consuming a strict gluten-free diet. *The American Journal of Clinical Nutrition, 72*(4), 937–939.

Barton, S. H., Kelly, D. G., & Murray, J. A. (2007). Nutritional deficiencies in celiac disease. *Gastroenterology Clinics of North America, 36*(1), 93–108.

Belay, G. (2009). Seed size effect on grain weight and agronomic performance of tef. *African Journal of Agricultural Research*, 836–839.

Berrios, J. D. J. (2010). Carbohydrate composition of raw and extruded pulse flours. *Food Research International*, 531–536.

Bonafaccia, G., Marocchini, M., & Kreft, I. (2003). Composition and technological properties of the flour and bran from common and tartary buckwheat. *Food Chemistry*, 9–15.

Boye, J., Zare, F., & Pletch, A. (2010). Pulse proteins: Processing, characterization, functional properties and applications in food and feed. *Food Research International*, 414–431.

Bressani, R. (1994). Composition and nutritional properties of amaranth. *Amaranth-Biology, Chemistry and Technology*, 185–205.

Brito, I. L. (2015). Nutritional and sensory characteristics of gluten-free quinoa (Chenopodium quinoa Willd)-based cookies development using an experimental mixture design. *Journal of Food Science and Technology*, 5866–5873.

Bruni, R. (2001). Wild Amaranthus caudatus seed oil, a nutraceutical resource from Ecuadorian flora. *Journal of Agricultural and Food Chemistry*, 5455–5460.

Bultosa, G., Teff. (2016). *Overview*.

Campos-Vega, R., Loarca-Piña, G., & Oomah, B. D. (2010). Minor components of pulses and their potential impact on human health. *Food Research International*, 461–482.

Caperuto, L. C., Amaya-Farfan, J., & Camargo, C. R. O. (2001). Performance of quinoa (Chenopodium quinoa Willd) flour in the manufacture of gluten-free spaghetti. *Journal of the Science of Food and Agriculture*, 95–101.

Capriles, V. D., Fernanda, G.d. S., & José Alfredo, G. A.a. (2016). Gluten-free breadmaking: Improving nutritional and bioactive compounds. *Journal of Cereal Science*, 83–91.

Carcea, M. (2020). Nutritional value of grain-based foods. *Food*, 504–506.

Chauhan, A., Saxena, D., & Singh, S. (2015). Total dietary fibre and antioxidant activity of gluten free cookies made from raw and germinated amaranth (Amaranthus spp.) flour. *LWT- Food Science and Technology*, 939–945.

Chethan, S., & Malleshi, N. (2007). Finger millet polyphenols: Optimization of extraction and the effect of pH on their stability. *Food Chemistry*, 862–870.

Chillo, S. (2009). Properties of quinoa and oat spaghetti loaded with carboxymethylcellulose sodium salt and pregelatinized starch as structuring agents. *Carbohydrate Polymers*, 932–937.

Correia, I. (2010). Comparison of the effects induced by different processing methods on sorghum proteins. *Journal of Cereal Science, 51*(1), 146–151.

Crockett, R., Ie, P., & Vodovotz, Y. (2011). Effects of soy protein isolate and egg white solids on the physicochemical properties of gluten-free bread. *Food Chemistry*, 84–91.

Dahele, A., & Ghosh, S. (2001). Vitamin B12 deficiency in untreated celiac disease. *The American Journal of Gastroenterology*, 745–750.

Demirkesen, I. (2010). Utilization of chestnut flour in gluten-free bread formulations. *Journal of Food Engineering*, 329–336.

Di Cairano, M. (2018). Focus on gluten free biscuits: Ingredients and issues. *Trends in Food Science and Technology*, 203–212.

Dietrych-Szostak, D., & Oleszek, W. (1999). Effect of processing on the flavonoid content in Buckwheat grain. *Journal of Agricultural and Food Chemistry*, 4384–4387.

Dini, I., Tenore, G. C., & Dini, A. (2004). Phenolic constituents of Kancolla seeds. *Food Chemistry*, 163–168.

do Nascimento, A. B. (2014). Availability, cost and nutritional composition of gluten-free products. *British Food Journal.*

Drabińska, N., Zieliński, H., & Krupa-Kozak, U. (2016). Technological benefits of inulin-type fructans application in gluten-free products–A review. *Trends in Food Science and Technology*, 149–157.

Drzewiecki, J. (2003). Identification and differences of total proteins and their soluble fractions in some pseudocereals based on electrophoretic patterns. *Journal of Agricultural and Food Chemistry*, 7798–7804.

Dura, A., & Rosell, C. M. (2016). Physico-chemical properties of corn starch modified with cyclodextrin glycosyltransferase. *International Journal of Biological Macromolecules*, 466–472.

Duta, D. E., & Culetu, A. (2015). Evaluation of rheological, physicochemical, thermal, mechanical and sensory properties of oat-based gluten free cookies. *Journal of Food Engineering*, 1–8.

El-Alfy, T. S., Ezzat, S. M., & Sleem, A. A. (2012). Chemical and biological study of the seeds of Eragrostis tef (Zucc.) Trotter. *Natural Product Research*, 619–629.

Ferreira, S. M. R. (2016). Utilization of sorghum, rice, corn flours with potato starch for the preparation of gluten-free pasta. *Food Chemistry*, 147–151.

Fiorda, F. A. (2013). Microestructure, texture and colour of gluten-free pasta made with amaranth flour, cassava starch and cassava bagasse. *LWT- Food Science and Technology*, 132–138.

Flores-Silva, P. C., Rodriguez-Ambriz, S. L., & Bello-Pérez, L. A. (2015). Gluten-free snacks using plantain–chickpea and maize blend: chemical composition, starch digestibility, and predicted glycemic index. *Journal of Food Science*, C961–C966.

Foschia, M. (2016). Nutritional therapy – Facing the gap between coeliac disease and gluten-free food. *International Journal of Food Microbiology*, 113–124.

Furlán, L. T. R., Padilla, A. P., & Campderrós, M. E. (2015). Improvement of gluten-free bread properties by the incorporation of bovine plasma proteins and different saccharides into the matrix. *Food Chemistry*, 257–264.

Gallagher, E., Gormley, T. R., & Arendt, E. K. (2014). Recent advances in the formulation of gluten-free cereal-based products. *Trends in Food Science & Technology*, 143–152.

Gebremariam, M. M., Zarnkow, M., & Becker, T. (2014). Teff (*Eragrostis tef*) as a raw material for malting, brewing and manufacturing of gluten-free foods and beverages: A review. *Journal of Food Science and Technology*, 2881–2895.

Giacco, R., Vitale, M., & Riccardi, G. (2016). Pasta: Role in diet. *The Encyclopedia of Food and Health*, 242–245.

Giménez, M. A. (2016). Nutritional improvement of corn pasta-like product with broad bean and quinoa. *Food Chemistry*, 150–156.

Giuberti, G., & Gallo, A. (2018). Reducing the glycaemic index and increasing the slowly digestible starch content in gluten-free cereal-based foods: A review. *International Journal of Food Science & Technology*, 50–60.

Gorinstein, S. (2002). Characterisation of pseudocereal and cereal proteins by protein and amino acid analyses. *Journal of the Science of Food and Agriculture*, 886–891.

Gungormusler, M., Basınhan, I., & Uçtug, F. G. (2007). Optimum formulation determination and carbon footprint analysis of a novel gluten-free pasta recipe using buckwheat, teff, and chickpea flours. *Journal of Food Processing and Preservation.*

Hager, A. S. (2013). Starch properties, in vitro digestibility and sensory evaluation of fresh egg pasta produced from oat, teff and wheat flour. *Journal of Cereal Science*, 156–163.

Hallert, C. (2002). Evidence of poor vitamin status in coeliac patients on a gluten-free diet for 10 years. *Aliment Pharmacology*, 1333–1339.
Hegde, P. S., Rajasekaran, N. S., & Chandra, T. (2005). Effects of the antioxidant properties of millet species on oxidative stress and glycemic status in alloxan-induced rats. *Nutrition Research*, 1109–1120.
Jnawali, P., Kumar, V., & Tanwar, B. (2016). Celiac disease: Overview and considerations for development of gluten-free foods. *Food Science and Human Wellness*, 169–176.
Kadan. (2001). Texture and other physicochemical properties of whole rice bread. *Journal of Food Science*, 940–944.
Kalinova, J., & Moudry, J. (2006). Content and quality of protein in proso millet (Panicum miliaceum L.) varieties. *Plant Foods for Human Nutrition*, 43–47.
Kaur, M. (2015). Gluten free biscuits prepared from buckwheat flour by incorporation of various gums: Physicochemical and sensory properties. *LWT- Food Science and Technology*, 628–632.
Klimczak, I., Małecka, M., & Pachołek, B. (2002). Antioxidant activity of ethanolic extracts of amaranth seeds. *Food/Nahrung*, 184–186.
Kumar, A. (2018). Millets: A solution to agrarian and nutritional challenges. *Agriculture & Food Security*, 31–33.
Kumar, C. M. (2019). Effect of incorporation of sodium caseinate, whey protein concentrate and transglutaminase on the properties of depigmented pearl millet based gluten free pasta. *LWT*, 19–26.
Laleg, K. (2016). *Structural, culinary, nutritional and anti-nutritional properties of high protein, gluten free, 100% legume pasta.*
Lasa, A. (2017). Nutritional and sensorial aspects of gluten-free products. *Nutritional and Analytical Approaches of Gluten-Free Diet in Celiac Disease*, 59–78.
Lee, A. (2009). The effect of substituting alternative grains in the diet on the nutritional profile of the gluten-free diet. *Journal of Human Nutrition and Dietetics, 22*(4), 359–363.
Lissner, L., & Heitmann, B. L. (1995). Dietary fat and obesity: Evidence from epidemiology. *European Journal of Clinical Nutrition, 49*(2), 79–90.
Maaran, S. (2014). Composition, structure, morphology and physicochemical properties of lablab bean, navy bean, rice bean, tepary bean and velvet bean starches. *Food Chemistry*, 491–499.
Marco, C., & Rosell, C. M. (2008). Breadmaking performance of protein enriched, gluten-free breads. *European Food Research and Technology*, 1205–1213.
Marengo, M. (2015). Molecular features of fermented and sprouted sorghum flours relate to their suitability as components of enriched gluten-free pasta. *LWT- Food Science and Technology*, 511–518.
Marti, A., & Pagani, M. A. (2013). What can play the role of gluten in gluten free pasta? *Trends in Food Science and Technology, 31*(1), 63–71.
Martínez, M. L. (2017). Effect of defatted almond flour on cooking, chemical and sensorial properties of gluten-free fresh pasta. *International Journal of Food Science & Technology*, 2148–2155.
Mastromatteo, M. (2011). Formulation optimisation of gluten-free functional spaghetti based on quinoa, maize and soy flours. *International Journal of Food Science & Technology*, 1201–1208.
Mathanghi, S. (2012). *Functional and phytochemical properties of finger millet (Eleusine coracana L.) for health.*
Matos, M. E., & Rosell, C. M. (2015). Understanding gluten-free dough for reaching breads with physical quality and nutritional balance. *Journal of Science and Food Agriculture*, 653–661.
McCrory, M. A. (2010). Pulse consumption, satiety, and weight management. *Advances in Nutrition*, 17–30.
McDonough, C. M., Lloyd, W., & Serna-Saldivar, S. (2000). *The millets: Handbook of cereal science and technology*. CRC Press.
Melini, F. (2017). Current and forward-looking approaches to technological and nutritional improvements of gluten-free bread with legume flours: A critical review. *Comprehensive Reviews in Food Science and Food Safety*, 1101–1122.

Molinari, R. (2018). Tartary buckwheat malt as ingredient of gluten-free cookies. *Journal of Cereal Science*, 37–43.

Molina-Rosell, C., & Matos, M. E. (2015). Market and nutrition issues of gluten-free foodstuff. *OmniaScience Monographs.*

Naqash, F. (2017). Gluten-free baking: Combating the challenges-A review. *Trends in Food Science and Technology*, 98–107.

Nunes, M. H. B., Ryan, L., & Arendt, E. K. (2009). Effect of low lactose dairy powder addition on the properties of gluten-free batters and bread quality. *European Food Research and Technology*, 31–41.

Ojetti, V. (2005). High prevalence of celiac disease in patients with lactose intolerance. *Digestion*, 106–110.

Omoba, O. S., Taylor, J. R., & de Kock, H. L. (2015). Sensory and nutritive profiles of biscuits from whole grain sorghum and pearl millet plus soya flour with and without sourdough fermentation. *International Journal of Food Science & Technology*, 2554–2561.

Oomah, B. (2011). *Pulse foods: Processing, quality and nutraceutical applications.*

Oomah, B. D., & Mazza, G. (1996). Flavonoids and antioxidative activities in buckwheat. *Journal of Agricultural and Food Chemistry*, 1746–1750.

Rachman, A. (2020). Gluten-free pasta production from banana and cassava flours with egg white protein and soy protein addition. *International Journal of Food Science and Technology.*

Ragaee, S., Abdel-Aal, E.-S. M., & Noaman, M. (2006). Antioxidant activity and nutrient composition of selected cereals for food use. *Food Chemistry, 98*(1), 32–38.

Rani, P. (2018). Impact of fermentation and extrusion processing on physicochemical, sensory and bioactive properties of rice-black gram mixed flour. *LWT*, 155–163.

Rocchetti, G. (2017), Impact of boiling on free and bound phenolic profile and antioxidant activity of commercial gluten-free pasta. Food Research International 69-77.

Rocchetti, G., Giuberti, G., & Lucini, L. (2018). Gluten-free cereal-based food products: The potential of metabolomics to investigate changes in phenolics profile and their in vitro bioaccessibility. *Current Opinion in Food Science*, 1–8.

Roosjen, J. (2005). *Processing of Teff Flour.* European patent specification, publication number WO. 2005.

Roy, F., Boye, J., & Simpson, B. (2010). Bioactive proteins and peptides in pulse crops: Pea, chickpea and lentil. *Food Research International*, 432–442.

Różyło, R. (2015). Effect of adding fresh and freeze-dried buckwheat sourdough on gluten-free bread quality. *International Journal of Food Science & Technology*, 313–322.

Ruales, J., & Nair, B. M. (1993). Content of fat, vitamins and minerals in quinoa (Chenopodium quinoa, Willd) seeds. *Food Chemistry*, 131–136.

Rybicka, I., & Gliszczyńska-Swig, A. (2017). Minerals in grain gluten-free products. The content of calcium, potassium, magnesium, sodium, copper, iron, manganese, and zinc. *Journal of Food Composition and Analysis*, 61–67.

Rybicka, I., Doba, K., & Bińczak, O. (2019). Improving the sensory and nutritional value of gluten-free bread. *International Journal of Food Science & Technology*, 2661–2667.

Saleh, A. S. M. (2013). Millet grains: Nutritional quality, processing, and potential health benefits. *Comprehensive Reviews in Food Science and Food Safety, 12*(3), 281–295.

Scazzina, F. (2015). Glycaemic index of some commercial gluten-free foods. *European Journal of Nutrition*, 1021–1026.

Scazzina, F., Siebenhandl-Ehn, S., & Pellegrini, N. (2013). The effect of dietary fibre on reducing the glycaemic index of bread. *British Journal of Nutrition*, 1163–1174.

Schoenlechner, R. (2010). Effect of water, albumen and fat on the quality of gluten-free bread containing amaranth. *International Journal of Food Science & Technology*, 661–669.

Sedej, I. (2011). Quality assessment of gluten-free crackers based on buckwheat flour. *LWT- Food Science and Technology*, 694–699.

Segura, M. E. M., & Rosell, C. M. (2011). Chemical composition and starch digestibility of different gluten-free breads. *Plant Foods for Human Nutrition*, 224–225.

Sharma, S., Saxena, D. C., & Riar, C. S. (2016). Nutritional, sensory and in-vitro antioxidant characteristics of gluten free cookies prepared from flour blends of minor millets. *Journal of Cereal Science*, 153–161.

Singh, K., Mishra, A., & Mishra, H. (2012). Fuzzy analysis of sensory attributes of bread prepared from millet-based composite flours. *LWT- Food Science and Technology*, 276–282.

Sparvoli, F. (2016). Exploitation of common bean flours with low antinutrient content for making nutritionally enhanced biscuits. *Frontiers in Plant Science*, 928.

Sripriya, G., Antony, U., & Chandra, T. (1997). Changes in carbohydrate, free amino acids, organic acids, phytate and HCl extractability of minerals during germination and fermentation of finger millet (Eleusine coracana). *Food Chemistry*, 345–350.

Standard, C., & Codex Alimentarius Commission. (2007). *Draft revised codex standard for foods for special dietary use for persons intolerant to gluten Joint FAO/WHO Food Standards Programme*. WHO.

Stantiall, S. E., & Serventi, L. (2018). Nutritional and sensory challenges of gluten-free bakery products: A review. *International Journal of Food Sciences and Nutrition*, 427–436.

Steadman, K. J. (2000). Fagopyritols, D-chiro-inositol, and other soluble carbohydrates in buckwheat seed milling fractions. *Journal of Agricultural and Food Chemistry*, 2843–2847.

Suliburska, J. (2013). Evaluation of the content and the potential bioavailability of minerals from gluten-free products. Acta Scientiarum Polonorum. *Technologia Alimentaria*, 75–79.

Taghdir, M. (2017). Effect of soy flour on nutritional, physicochemical, and sensory characteristics of gluten-free bread. *Food Science & Nutrition*, 439–445.

Tatham, A. (1996). Characterisation of the major prolamins of tef (Eragrostis tef) and finger millet (Eleusine coracana). *Journal of Cereal Science*, 65–71.

Taylor, J., & Parker, M. (2002). Quinoa. pseudocereals and less common cereals: Grain properties and utilization potential. 100–101.

Thompson, T. (2000). Folate, iron, and dietary fiber contents of the gluten-free diet. *Journal of the Academy of Nutrition and Dietetics, 1389*.

Tikkakoski, S., Savilahti, E., & Kolho, K.-L. (2007). Undiagnosed coeliac disease and nutritional deficiencies in adults screened in primary health care. *Scandinavian Journal of Gastroenterology, 42*(1), 60–65.

Tiwari, B. K., & Singh, N. (2012). *Pulse chemistry and technology*. Royal Society of Chemistry.

Vici, G. (2016). Gluten free diet and nutrient deficiencies: A review. *Clinical Nutrition*, 1236–1241.

Wolter, A. (2014). Influence of sourdough on in vitro starch digestibility and predicted glycemic indices of gluten-free breads. *Food & Function*, 564–572.

Zandonadi, R. P. (2012). Green banana pasta: an alternative for gluten-free diets. *Journal of the Academy of Nutrition and Dietetics*, 1068–1072.

Zhu, F. (2018). Chemical composition and food uses of teff (Eragrostis tef). *Food Chemistry*, 402–415.

# Chapter 3
# Gluten-Free Food: Role of Starch

**Sandeep Singh Rana and Payel Ghosh**

**Abstract** A rising demand for gluten-free foods is triggered by growing cases of celiac disease, but also by a trend towards removing all potentially allergenic proteins in a diet. It's a known fact that gluten elimination impacts the product structure and texture significantly. It is difficult to alter a gluten-free product recipe that would offer a product comparable to conventional food. One of the key components of the gluten-free product is the starch of a specific botanical origin. Additionally, their properties may be changed by compatible shape and texture-forming ingredients or additives, including multiple texturizing aids. The function of starch is often significant in these structures, as its proper choice and Treatment may have a direct impact on the finished products. An evaluation of the literature identifies starch as a key component in gluten-free food items. This starch structure shows variations between different forms of this biopolymer and their effect on the characteristics of the goods.

**Keywords** Starch · Gluten-free product · Texture

## 3.1 Introduction

Increasing competition for GF foods is linked to a rising number of patients living with the gluten-free diet. Some illnesses, about which the removal of gluten resulted in better wellbeing, have contributed to a trend of removing it off the diet of healthier people having a strong effect on the GF industry. It is found that 20% of US customers buy GF items (Nijeboer et al., 2013). According to analyses, 65% of customers purchase these items as they find them to be safe, 27% for weight loss, 11% for health benefits) and remaining for other benefits (Watson et al., 2014). The total size of the GF industry in the world was to be $6.2 billion in 2018, of which 59% is in the USA alone. According to EU regulations, gluten origins include

S. S. Rana (✉) · P. Ghosh
Department of Food Technology, Vignan's Foundation for Science Technology and Research, Vadlamudi, Andhra Pradesh, India



N. Singh Deora et al. (eds.), *Challenges and Potential Solutions in Gluten Free Product Development*, Food Engineering Series,
https://doi.org/10.1007/978-3-030-88697-4_3

'wheat (such as spelled and khorasan wheat), rye, barley, oats or their hybridized varieties, and items thereof,' with certain variations.

According to Regulation pertaining to commission (EC No. 41/2009) items including 'ingredients produced of wheat, rye, barley, oats or their cross-breeding variants that had specifically formulated to minimize gluten' and containing gluten content lower than 100 ppm per kilogram may be referred to as 'very low gluten. The items called "gluten-free" did not achieve more than 20 mg of gluten per lb. The US Food and Drug Administration introduced a similar definition that restricts the amount of gluten in GF items to 20 ppm (Grain Labeling; Gluten-Free Marking of Fiorda et al., 2013a). Within these restrictions, starches (corn, potatoes, cassava, rice) are the major raw materials that may be used in the processing of Gluten-Free food (Deora et al., 2014, 2015; Brito et al. 2014; Fiorda et al. 2013b; de la Hera et al. 2014; Korus et al., 2015b). A variety of reports have lately been conducted on the use of hydrocolloids, dietary nutrients, additives, and manufacturing aids appropriate for these processes. (Deora et al., 2014, 2015; Korus et al. 2015a; Marco and Rosell, 2008).

## 3.2 Properties of Starch

In several plants, starch is important storage polymers. It is composed of two groups of proteins, spiraled amylopectin and linear amylose. In both cases, the building block is the P-D-glucopyranose residue, which forms alpha-1,4-glucosidic bonds in linear amylose structure and additional P-1,6-glycosidic branches in amylopectin molecules. Distinctions in the framework of both polymers result insignificant. Amylose seems to be more resistant to crystallization, or retrograde, whereas amylopectin could be spread in liquid but is much slower, resulting in soft gels and poor films (Pérez and Bertoft, 2010). Indigenous starch is arranged in granules found in plants and also in stems. They range in scale (0.1–200 μm) and form is related to the botanical roots. The partial crystalline structure of starch influences its properties. This structure is also of paramount importance for the other components apart from starch which is often used in food processing and other industries. In addition to direct alteration, starch structures may be changed by the careful collection of components, often hydrocolloids, reacting with stabilizing starch.

## 3.3 Flour

Flours typically consisting of only starch can mostly be utilized for the GF product development as a substitute for wheat flour. Starch is the main component that specifically influences the properties of starch, beginning with its outward presence. Their inclusion in flours greatly influences the profile of nutrition, for example, starch-based bread that produces many times less protein than comparable products with the inclusion of buckwheat or millet flours. The milling operation may modify the intrinsic behavior of starch. For example, mechanical shearing is known to impact the behaviors of starch in GF products and thus it becomes important to understand the aspect of milling for GF application.

## 3.4 Function of Starch in GF Products

The addition of moisture has a considerable on the properties of the starch. For example, the viscoelastic behavior, as well as the formation of continuous bonds in the products, is greatly influenced by the starch properties. The precise relationship of these structures with GF products depends heavily on the botanical origins as well as other chemical properties of which can participate in its stabilization. Water causes starch granules to swell, depending on the origin of starch. Tegge (2004) suggests that starches in terms of water absorption potential may be classified into three classes. The heat stability of starch-water systems causes granules to swell, thus increasing volume and altering other characters. At a considerably lower level of humidity ends in a modification of the crystalline structure, morphology, chemical and enzyme sensitivity of the granules (Hoover, 2010).

The higher-water cycle is called rinsing, which often contributes to mechanical properties being modified (Tester et al., 1998). The starch swelling yields a more compressed form as thermal treatment of resultant starch commences and mixed with a minute volume of water that is shown as a consequence of growing resistance to the mixing during treatment which involves thermal as well as mechanical energy. Granules growing to disintegrate over those levels. The exact transition parameters are related to the type of starch granules. For example, structural type, granular behavior, water level, rate of heating. Amylose leak can also be understood by the viscosity analysis. In the end stage, starch granulate break, contributing to a transient decrease in paste viscosity and enzyme sensitivity (Wang & Copeland, 2013). The early phases of the thermal starch transformation in the water are termed gelatinization, whereas the latter are termed gelatinization. However, the use of these conditions among scientists is not consistent (Xie et al., 2005). It should be noted that when you cool down the starch, 3D Structure which can attach sizeable quantities of water meritoriously (Singh et al., 2003). Active cooling contributes to gel-forming (Pycia et al., 2012). During food storage, the mechanism of accumulation of the starch collectively termed retrogradation, has a major effect on certain characteristics of foodstuffs (Delcour et al., 2010). Reconnection of molecules of amylopectin via the chemical bonds of hydrogen alters the textural properties of starch gels. Acceleration can be done by freezing cycles whereby water molecules are separated from starch additives and hydrogen bonds can be left empty.

Different forms of native starch vary in the propensity to retrogradation, this phase is inevitable and typically contributes to food quality degradation. In large part, this describes the extensive usage in different sectors of the food sector. Starch in storage cells has no impact on plant tissue properties and is not essential to the digestion that produces the first responsive starch fractions. During the process of mastication, fragile cell walls can be certainly damaged, the most amylolytic enzymes, for example, banana starch, cannot still have starch granules. After physical or chemical treatment, the role of starch in delivering physical starch characteristics is increasing. The composition and presence of gelatinized starch, for example, in potato tuber effects, is the origin of the potato classification (Kaur et al., 2002). The impact of the thermo-mechanical application on starchy can affect certain granules and alter their surface characteristics which have direct effects on other properties that impact the final product. The magnitude of these improvements may be managed to a large degree by the correct selection of input material.

The varying degree of amylose content in the cornmeal is known to impact the textual behavior of tortillas and snacks'. Snack from extrusion also requires the correct choice of raw materials. The expansion and quality of the goods should be satisfactory. Type of the Starch has an important aspect in the development, as its extent depends directly on the material viscosity. The gas will not be caught if it is too small, which contributes to an unexpanded error. Expansion is constrained if it is too powerful (Guy, 2001; Moraru & Kokini, 2003). Cereal having undergone the operation of fine milling has the presence of a higher surface area. This higher degree of the surface area creates more water adsorption and serves as the adhesive for flour pieces along with hydrocolloids, enabling them to establish a clear structural framework of a bread. Starch and water can not shape a dough alone with sufficient mechanical properties, but the inclusion of even limited quantities of soluble polysaccharides or proteins enables the creation of a network that is Visco-elastic and can be modeled and drained in various ways. While the key element influencing its consistency, conventional pasta is made of wheat durum semolina, there are several comparable items dependent on maize. The consistency of starch is especially relevant in such situations. The above applies in particular to certain Asian noodles, in which the dough is made exclusively from native starch and starch paste, which is formulated from about 5% of the total starch (Tan et al., 2009).

Bread is one of the main foodstuffs. Traditionally, the wheat meal may also be used as food from other grains including rye, rice, or maize. The development of GF bread is very different as compared to traditional products derived from wheat. Since gluten protein is absent, a long mixture, which is usually necessary for developing the protein network, is not necessary for the dough. There can also be improved water supplements and the usage of batter-like formulae in GF pastry output is popular. An introduction of enzymes, emulsifiers, and dietary supplements may bring more modifications. Starch gelatination is important for baking the food. The variety, in oven conditions, is based on the form of starch: crystalline composition, amylose material, size, and shape of the granule. Other dough formulation components could affect the behavior of starch, particularly by altering water availability. If the gelatinization decreases, the process can be carried out faster when more water is freed from other dough components after heat. This may in turn influence the cooling of gelling and retrograde starch.

The function of stärch is generally linked to their viscosity in these low-concentration systems. Many of those goods are focused on natural starchy goods, but those based on modified starches are the most significant industrial items. The usage of adjusted starches makes for a quicker processing of food, greater viscosity regulation, and increased product consistency. While it is not appropriate to attach starch to goods that are built upon certain structural biopolymers including animal protein, as in many conventional products, it can be used as a bulking agent or development assist which may avoid water losses. Due to its nature, starch can be used as a plastic substitute. It involves the production of comestible films and coatings that may be used in specific food divisions (Bertolini, 2009). Starch films may be used to prolong the shelf life of the drug, to enhance its quality, and to avoid the mechanical loss or a combination of separate food components. Starch is often used in jelly development as a simple, recyclable substitute for molds (Radley, 1976).

## 3.5 Role of Starch in the GF Products

The absence of Gluten makes the role of statch important for the structure building. GF texture and structure Products, please. Corn, wheat, tapioca, rice, and rice are the most important sources of starch in these products. Potatoes. Unquestionable (Capriles & Arêas, 2014). Sorghum, millet, or tef as well as pseudocereals, such as corn quinoa and buckwheat, are normally used in molten form while corn and rice are widely used as insulated stomach and flour. Roots, tubers, vegetables, and other sources, typically as components of composite mixes may also be used. These constituents can be used as an admixture to boost texture, sensory qualities, and the nutritional value of GF products.

Gluten deficiency contributes instead to liquid batters and can lead to baked bread with cracking structures, bad color, and other defects in postbaking consistency. The lower amounts of these substances, such as protein, vitamins, minerals, and dietary fiber, have a greater nutritional value compared with gluten-based foods fiber.

There are two approaches to the study. Gluten is being modified to eliminate its toxicity, and gluten-deprived recipes are becoming designed to suit better sensor and nutritional benefit criteria By fortifying with isolated compounds or incorporating natural raw materials rich in nutritionally useful ingredients, GF's nutritional benefit is increased. Additional GF products with nutrients sometimes contribute to improvements in the physical-chemical properties of the drug which need extra precautions during industrial implementation. The attributes of starch-based end and intermediate items are primarily dictated by the starch characteristics of the recipe. The starch granules' relative size which relies on technologies for meal milling or starch insulation often affects the product's properties. The source of starch/meal and the equipment utilized influences granularity, bulking consistency, and exposure to starch. It affects the manufacturing efficiency of flowering and defines hydrating and properties of pasting (Abebe et al., 2015). The distribution by the scale of flour particles affects the characteristics of GF rice cupcakes, according to Kim and Shin (2014).

The volume of weakened stutter granules decreases with a decline in particle size, contributing to decreased water-binding capacitance, solubility, and lightweight, whereas rough proteins and yellowness thus decline. When the particle size decreases, the final and reverse viscosities of the rice meal improve. With a given quantity of cupcakes, the same pattern can be found when their strength and springiness decline as a particle size increases. As particle size increases with homogeneity, the air cell sizes decrease. Amylose leaching during starch gelatinization, according to the scientists, supports building network structure with the egg-white protein spray. It can also be found that smaller particles (high starch fractions) in rice cupcakes can develop tiny air cells after baking. As per de la Hera et al. (2013) decreased rice flour particle size initially affects optimistic basic bread length, but bad breadmaking efficiency is the best grain. de la Hera et al. (2013) Based on the depth of the stream, the strongest outcomes are reached for varying particle dimensions. Increases in the basic volume lead to GF commodity texture parameters.

As per de la Hera et al. (2013) the compound granules of the starch partially dissolve during kneading and are combined with protein, moisture, and Hydroxypropyl methylcellulose, to cover starch with relatively larger. Dough formation is influenced by the degree of this disintegration. The strongest baking outcomes may be achieved when soft rice varieties are used for flour development utilizing red-shaped Starch granules (Kang et al., 2015). Specific analysis on maize-based bread found fine flours in CMC systems was low and produced relatively smaller loaves in comparison to the with coarser particles. However, for the finished breast consistency, the function of the maize and method of milling appeared more significant.

For the quality of bread, it is necessary to process starch or starchy content before making the products. The findings demonstrate clearly that the properties of starch granules have a significant effect on manufacturing and finished products' consistency.

## 3.6 Source of Other Starch

The poor nutritional benefit of GF items promotes enhancement work. The substitution of essential constituents with nutritionally useful raw materials in such formulations is one of the techniques. In certain situations though, associations with starch and other device components enhance sensory conditions and increase the product's shelf-life. Very commonly, nutrition performance enhancement has been followed by structural degradation and texture degradation, and the work aims at seeking an optimal degree of addition to a specific ingredient. The suitability of such GF-products for processing has been tested for a collection with flours with specific botanical origins. They contain pseudocereal meal and pseudocereal meal (buckwheat, amaranth, quinoa), root, and tuberous meal (cassava, yam). Inputs from plants like cotton, leather seed, may be collected from flosses and used as ingredients.

## 3.7 Modified Starches in GF Products

In addition to the natural food portion of native starch, GF items that include chemically engineered starches, labeled as a food additive and labeled with "E" numbers. Due to its developed, durable properties, changed starches can be easily added to various food systems, without any noticeable degradation of the finished product, and for the replacement of wheat starches in conventional bakers up to 20%. According to Miyazaki et al. (2006), its implementation affects water absorption and the rheological properties of the flour, the degree of the gelatination of starches, the stiffness, and the staling of the bread.

Also studied in GF goods were modified starches. The rise in the bread amount was reported owing to systemic improvements correlated with the introduction of ADA and HDP. The parameter value of the texture was identical to the unit. The existence of chemical enhanced starch stabilizes the composition of the crumb and decreases the retrogradation of dissolved starch glucans, and therefore prevents stalking. The inclusion of cross-linked higher bread hardness may, on the other, be undesirable (Ziobro et al., 2012; Witczak et al., 2012).

In the processing of GF goods chemically adjusted starches may also be used. In comparison to the previous community it is known as food components (like native starch) and ultimately offers the user a 'safe name.' The effect of hydrothermally adjusted bean starch on GF bread has been studied by Krupa et al. (2010). We find that the usage of this component decreases crumb stiffness and retrogradation enthalpy substantially. Pregelatinized tapioca starch was used as a structural shaping agent in rice meat-based bread by Pongjaruvat et al. (2014). The existence of batter-like teats made the processing of the teeth less prone to shear. With the application of Chemical, enzymatic or physical techniques, maltodextrins produced may maintain bread texture. They are well-known for their role in delaying bread stalking. The application of low molecular weight dextrin to wheat bread decreases the retrogradation of starch, according to Miyazaki et al. (2004) but is not generally related to a reduction in bread hardiness. In experiments on starch gels' model systems, it has been shown that the existence of dextrins increases the gelatinization of starch and therefore decreases retrograde enthalpy (Durán et al., 2001).

Meal with medium-DE maltodextrins was observed having more water absorption and dough expansion relative to samples with low-DE maltodextrins. The bread volume made from flours having a marginal amylolytic activity, particularly with medium DE, was affected positively by maltodextrins. The partial substitution of starch bases (grain and potato) with observable DE maltodextrins is an improved gelatinization temperature and decreased viscosity of the pastes, according to Witczak et al. (2010). The maltodextrins incorporation has also been found to have aimpact dough rheological properties. Adrop in bread volume and weak bread consistency, marked by irregular pore shape and thickness is also reported due the presence of maltodextrins with small DE. The advent of maltodextrins with shorter chains affected the amount of bread and slow bread stalking. The maximum DE maltodextrins significantly limit recrystallation of amylopectin. Maltodextrins with a high equivalent of dextrose are demonstrated to be used as an anti-staling agent for GF Broad manufacturing (Witczak et al., 2010). Ferreira et al. (2014), who applied enzymatically modified rice meal (via alpha-amylase) to baby food formulations, also conducted studies on hydrolyzed starch in the GF formulation.

## 3.8 Resistant Starches in GF Products

Growing understanding among consumers of food's dietary role generates augmented interest in role of starch-resistant enzymes in human dietary tract. Starch can be divided into fast (FDS) and slow -diagnostic starch (SDS). The non-digesting component, "strong starch" (RS, Birt et al., 2013; Fuentes-Zaragoza et al., 2011) of starch and of its partially hydrolysed derivatives that isn't absorbed in tiny gutes. This may also be distinguished in different categories. For example, Resistant starch type 1 is described as intact, physically inaccessible starch in tissues and cells. Resistant starch type 1 (RS1) Second category (RS2) is a natural, unhealthy starch produced by human amylases from other plants and third (RS3) is a retrograded starch generated while handling starch gels (e.g., bread). This is a non-gelatinised starch. Form 4 (RS 4) contains all chemically transformed starches that due to their altered composition can not be reached by the enzymes (Öztürk & Kökse, 2014). Resistant starch may have multiple essential functions and has a health-friendly quality. Compared to other dietary fiber fractions, it decreases pressure of food, enhances satiety and strengthens digestive functions (Fuentes-Zaragoza et al., 2011).

The involvement of food has a beneficial impact on absorption to glucose, which reduces the blood postprandial level (Birt et al., 2013). Resistant starch may then be fermented in the bowel after leaving small intestines, providing short chains of fatty acids, growing pH and controlling microflora production. Pre-biotic formation, production and interaction of bacteria in probiotic products are promoted and unnecessary microorganisms are removed or decreased (Birt, et al., 2013; Fuentes-Zaragoza et al., 2011). It is also reported that starch (resistant) can possibly bind bile acids and increase sterol excretion resulting in a decrease in blood plasma levels of cholesterol and triacyl glycerols (Birt et al., 2013; Öztürk & Kökse, 2014). Large and bakery items are among the most significant sources of starch in human diet and are thus supported by the usage of resistant starch. Wojciechowicz-Budzisz et al. (2015) examined the effects on consistency of wheat dough and bread of retrograded acetylated starch (RS4). Their addition to the formulation up to 10% did not substantially decrease bread quality.

The increased amount of resistant starch also has been observed to deteriorate meal quality and decrease bread output. The fractional substitution of flour with resistant amber preparations has been shown to have a major impact on the changes in rheology of the pudding and to increase its spring, in particular when 15% of the RS3 is applied. As a consequence, the technical procedure has to be modified and sufficient water to produce the dough of correct strength used in the formulae. The effect of wheat flour replacement with adjusted pea starch, providing high RS yield and bread consistency was examined by Sanz-Penella et al. (2010). It has been reported that the RS content improved dramatically, with no adverse effects on the rheology of dough and acceptability of the bread at quantities up to 20%. Therefore, the retrogradation rate of amylopectin was high. Research has also been performed on GF food to determine the effect of resistant starch on the consistency of starch-based items. The results of maize/potato starch substitution in GF bread formulation with maize and tapioca-resistant starch preparations were studied by Korus et al. (2009).

Bread with RS was defined in contrast to regulation by softer crumb. Specific studies also indicate that introducing chemically modified starch (increasing RS4 level) dramatically affects dough rheology, increases the consistency of bread and reduces crumb strength, and boosts its elasticity in quantities up to 15%. Upon applying high amylose maize starch, however, no beneficial results were found (Witczak et al., 2012; Ziobro et al., 2012; Tsatsaragkou et al., 2014).

The existence of starch (resistant) has no impact on the firmness of crumb but greatly enhances its elasticity, particularly at the introduction of 15% of rice meal. Adding carob flour could contribute to more improvements. The effect of RS preparation on GF cookies is dependant upon rice flour and tapioca starch (Tsatsaragkou et al., 2015). The incorporation improved elasticity and thinner batters of coke. The special amount of cakes decreased to 15% RS, as the degree of RS rose. Similar porosity, the number of pores, and an improvement in the average diameter of pore have been seen by crumb grain analysis with-RS concentration. In formulations with increased amounts of RS, cake crumb stayed weaker throughout the transport. Sensorial assessment of cakes shows that all formulas are approved, often with the cake that includes 20% RS. Ren and Shin (2013) investigated the impact of resistant starch (RS4) on the properties of dough and consistency of typical rice cookies. The findings indicated that the dietary fiber proportion of the final items was increased. In comparison, swelling strength, solubility, water binding capability, and all gluing viscosities decreased with an increase in RS4 material. The satisfaction check found that the additional RS4 color and overall consistency were the best when 10% of the RS4 was applied.

The inclusion of flours rich in this fraction may also contribute to the increase in resistant starch, and therefore dietary fiber, in GF items. The results rheology of dough and bread properties of the introduction of buckwheat flame by 30% and 50% were studied by Wronkowska and Soral-Śmietana (2008). The amount of amylase-resistant starch was found to be higher, apart from the growth of protein and minerals. Dough dilution and decrease in viscosity are induced by the addition of buckwheat meal. The bean meal was the source of resistant starch in GF spaghetti dependent on rice starch in another research (Giuberti et al., 2015). While after this supplement the overall starch and RDS and SDS fractions decreased (an improvement for protein, ash, and overall fiber has been found), it has been supplemented by substantial RS rises of around 30%. In the introduction, the introduction of flour was found to improve the optimal cooking period and water absorption without impacting cooking loss. Properties in shape. The Green Plantain Flour product used by Sarawong et al. (2014) was used for the manufacture of GF Bread from a combination of rice and GF Wheat Starch.

The findings suggest that such an improvement in quantities of up to 30% will have an appropriate consistency for finished goods and greatly increase the resistor starch content. Besides, the authors developed and refined technical parameters and applied water to achieve the full amount of bread and soft and porous butter.

## 3.9 Conclusion

In GF products, starch and its derivatives play a substantial function, like modifiable starches. Without gluten, the key structure and function part of several structures is starch. The usage of various flours contributes to the goods in nutrients: calcium, vitamins, and minerals. GF products may contain starch adjusted with chemical, physical or enzymatic treatment in addition to natural starch. We usually play a crucial role in enhancing the texture properties and avoiding harassment. Preparations for resistant starch boost GF crop nutritionally. In the GF industry, the high variability of starch types is still insufficient, leading to more research on recipes that contain customized starch mixes for GF products with high quality.

## References

Abebe, W., Collar, C., & Ronda, F. (2015). Impact of variety type and particle size distribution on starch enzymatic hydrolysis and functional properties of tef flours. *Carbohydrate Polymer*, 260–268.

Bertolini, A. C. (2009). Trends in starch applications. In A. C. Bertolini (Ed.), *Starchescharacterization, properties, and applications* (pp. 1–19). CRC Press.

Birt, D. F., Boylston, T., Hendrich, S., Jane, J. L., Hollis, J., Li, L., McClelland, J., Moore, S., Phillips, G. J., Rowling, M., Schalinske, K., Scott, M. P., & Whitley, E. M. (2013). Resistant starch: Promise for improving human health. *Advances in Nutrition*, 587–601.

Brito, I. L., de Souza, E. L., Felex, S. S. S., Madruga, M. S., Yamashita, F., & Magnani, M. (2014). Nutritionalandsensorycharacteristicsofgluten-free quinoa (*Chenopodiumquinoa* Willd). *Journal of Food Scienceand Technology*, 1–8.

Capriles, V. D., & Arêas, J. A. G. (2014). Novel approaches in gluten-free breadmaking: Interfacebetween food science, nutrition, and health. *Comprehensive Reviews in Food Science and Food Safety, 13*(5), 871–890.

de la Hera, E., Martinez, M., & Gómez, M. (2013). Influence of flour particle size on quality of gluten-free rice bread. *LWT-Food Science and Technology, 54*(1), 199–206.

de la Hera, E., Rosell, C. M., & Gomez, M. (2014). Effect of water content and flour particle size on gluten-free bread quality and digestibility. *Food Chemistry, 151*, 526–531.

Delcour, J. A., Bruneel, C., Derde, L. J., Gomand, S. V., Pareyt, B., Putseys, J. A., Wilderjans, E., & Lamberts, L. (2010). Fate of starch in food processing: From raw materials to finalfoodproducts. *Annual Review of Food Science and Technology, 1*, 87–111.

Deora, S., Deswal, A., & Mishra, H. N. (2014). Alternative approaches towards gluten-freedoughdevelopment: Recent trends. *Food Engineering Reviews*, 89–104.

Deora, S., Deswal, A., & Mishra, H. N. (2015). Functionality of alternative protein inglutenfreeproduct development. *Food Science and Technology International*, 364–379.
Durán, E., León, A., Barber, B., & de Barber, C. B. (2001). Effect of low molecular weight dextrins on gelatinization and retrogradation of starch. *European Food Research and Technology, 212*(2), 203–207.
Ferreira, S. M., Caliari, M., Júnior, M. S. S., & Beleia, A. D. P. (2014). Infant dairy-cereal mixture for the preparation of a gluten free cream using enzymatically modified rice flour. *LWT-Food Science and Technology*, 1033–1040.
Fiorda, F. A., Soares Júnior, M. S., da Silva, F. A., Souto, L. R. F., & Grosmann, M. V. E. (2013a). Amaranth flour, cassava starch and cassava bagasse in the production of gluten-free pasta technological and sensory aspects. *International Journal of Food Science & Technology*, 1977–1984.
Fiorda, F. A., Soares, M. S., da Silva, F. A., Grosmann, M. V. E., & Souto, L. R. F. (2013b). Microstructure, texture and colour of gluten-free pasta made with amaranth flour, cassava starch and cassava bagasse. *LWT-Food Science and Technology*, 132–138.
Fuentes-Zaragoza, E., Sánchez-Zapata, E., Sendra, E., Sayas, E., Navarro, C., Fernández-López, J., & Pérez-Alvarez, J. A. (2011). Resistant starch as prebiotic: A review. *Starch-Stärke, 63*(7), 406–415.
Giuberti, G., Gallo, A., Cerioli, C., Fortunati, P., & Masoero, F. (2015). Cooking quality andstarchdigestibility of gluten free pasta using new bean flour. *Food Chemistry*, 43–49.
Guy, R. (2001). Raw materials for extrusion cooking. In R. Guy (Ed.), *Extrusioncooking* (pp. 5–28). Technologies and application. Woodhead Publishing Ltd.
Hoover, R. (2010). The impact of heat-moisture treatment on molecular structures and properties of starches isolated from different botanical sources. *Critical Reviews in Food Science and Nutrition*, 835–847.
Kang, T.-Y., Sohn, K. H., Yoon, M.-R., Lee, J.-S., & Ko, S. (2015). Effect of the shape of rice starch granules on flour characteristics and gluten-free bread quality. *International Journal of Food Science and Technology*.
Kaur, L., Singh, N., Sodhi, N. S., & Gujral, H. S. (2002). Some properties of potatoes and their starches I. Cooking, textural and rheological properties of potatoes. *Food Chemistry*, 177–181.
Kim, J. M., & Shin, M. (2014). Effects of particle size distributions of rice flour on the quality ofgluten-free rice cupcakes. *LWT-Food Science and Technology*, 526–532.
Korus, J., Witczak, M., Ziobro, R., & Juszczak, L. (2009). The impact of resistant starch oncharacteristics of gluten-free dough and bread. *Food Hydrocolloids*, 988–995.
Korus, J., Witczak, T., Ziobro, R., & Juszczak, L. (2015a). Linseed (*Linum usitatissimum* L.) mucilage as a novel structure forming agent in gluten-free bread. *LWT- Food Science & Technology*, 257–264.
Korus, J., Witczak, M., Ziobro, R., & Juszczak, L. (2015b). The influence of acorn flour on rheological properties of gluten-free dough and physical characteristics of the bread. *European Food Research and Technology, 240*, 1135–1143.
Krupa, U., Rosell, C. M., Sadowska, J., & Soral-Śmietana, M. (2010). Bean starch as ingredient for gluten-free bread. *Journal of Food Processing and Preservation, 34*(2), 501–518.
Marco, C., & Rosell, C. M. (2008). Functional and rheological properties of protein enriched gluten free composite flours. *Journal of Food Engineering*, 94–103.
Miyazaki, M., Maeda, T., & Morita, N. (2004). Effect of various dextrin substitutions for wheat flour on dough properties and bread qualities. *Food Research International*, 59–65.
Miyazaki, M., Van Hung, P., Maeda, T., & Morita, N. (2006). Recent advances in application of modified starches for breadmaking. *Trends in food science & Technology, 17*(11), 591–599.
Moraru, C. I., & Kokini, J. L. (2003). Nucleation and expansion during extrusion and microwave heating of cereal foods. *Comprehensive reviews in food science and food safety, 2*(4), 147–165.
Nijeboer, P., Bontkes, H. J., Mulder, C. J., & Bouma, G. (2013). Non-celiac gluten sensitivity. Is it in the gluten or the grain? *Journal of Gastrointestinal & Liver Diseases, 22*(4).
Öztürk, S., & Kökse, H., (2014). Production and characterisation of resistant starch and its utilisation 798 as food ingredient: a review. *Quality Assurance and Safety of Crops & Foods, 6*(3), 335–346.

Pérez, S., & Bertoft, E. (2010). The molecular structures of starch components and their contribution to the architecture of starch granules: A comprehensive review. *Starch-Stärke, 62*(8), 389–420.

Pongjaruvat, W., Methacanon, P., Seetapan, N., Fuongfuchat, A., & Gamonpilas, C. (2014). Influence of pregelatinised tapioca starch and transglutaminase on dough rheology and quality of gluten-free jasmine rice breads. *Food Hydrocolloids, 36*, 143–150.

Pycia, K., Juszczak, L., Gałkowska, D., & Witczak, M. (2012). Physicochemical properties of starches obtained from Polish potato cultivars. *Starch-Stärke, 64*(2), 105–114.

Radley, J. A. (1976). Physical methods of characterising starch. *In Examination and analysis of starch and starch products* (pp. 91–131). Springer, Dordrecht.

Ren, C., & Shin, M. (2013). Effects of cross-linked resistant rice starch on the quality of Korean traditional rice cake. *Food Science and Biotechnology, 22*(3), 697–704.

Sanz-Penella, J. M., Wronkowska, M., Soral-Śmietana, M., Collar, C., & Haros, M. (2010). Impact of the addition of resistant starch from modified pea starch on dough and bread performance. *European Food Research and Technology, 231*(4), 499–508.

Sarawong, C., Schoenlechner, R., Sekiguchi, K., Berghofer, E., & Ng, P. K. (2014). Effect of extrusion cooking on the physicochemical properties, resistant starch, phenolic content and antioxidant capacities of green banana flour. *Food Chemistry, 143*, 33–39.

Singh, N., Singh, J., Kaur, L., Sodhi, N. S., & Gill, B. S. (2003). Morphological, thermal and rheological properties of starches from different botanical sources. *Food Chemistry, 81*(2), 219–231.

Tan, H. Z., Li, Z. G., & Tan, B. (2009). Starch noodles: History, classification, materials, processing, structure, nutrition, quality evaluating and improving. *Food Research International, 42*(5–6), 551–576.

Tegge, G. (2004). Stärke und Stärkederivate. 3. vollständig überarbeitete Auflage Hamburg: B. Behr's Verlag GmbH & Co. KG.

Tester, R. F., Debon, S. J. J., & Karkalas, J. (1998). Annealing of wheat starch. *Journal of Cereal Science, 28*(3), 259–272.

Tsatsaragkou, K., Gounaropoulos, G., & Mandala, I. (2014). Development of gluten free bread containing carob flour and resistant starch. *LWT-Food Science and Technology, 58*(1), 124–129.

Tsatsaragkou, K., Papantoniou, M., & Mandala, I. (2015). Rheological, physical, and sensory attributes of glutenfree rice cakes containing resistant starch. *Journal of Food Science, 80*(2), E341–E348.

Wang, S., & Copeland, L. (2013). Molecular disassembly of starch granules during gelatinization and its effect on starch digestibility: a review. *Food & function, 4*(11), 1564–1580.

Watson, M., Nasr, E. M., Hassan, M. S., Zamany, M. M., Ali, S. M., & Behnam, M. (2014). Evaluation of quality indicators related to quality bread wheat promising lines. *Russian Journal of Agricultural and Socio-Economic Sciences, 25*(1).

Witczak, M., Juszczak, L., Ziobro, R., & Korus, J. (2012). Influence of modified starches on properties of gluten-free dough and bread. Part I: Rheological and thermal properties of gluten-free dough. *Food Hydrocolloids, 28*(2), 353–360.

Witczak, M., Korus, J., Ziobro, R., & Juszczak, L. (2010). The effects of maltodextrins on gluten-free dough and quality of bread. *Journal of Food Engineering, 96*(2), 258–265.

Wojciechowicz-Budzisz, A., Gil, Z., Spychaj, R., Czaja, A., Pejcz, E., Czubaszek, A., & Zmijewski, M. (2015). Effect of acetylated retrograded starch (resistant starch RS4) on the nutritional value and microstructure of the crumb (SEM) of wheat bread. *Journal of Food Processing and Technology, 6*(6).

Wronkowska, M., & Soral-Smietana, M. (2008). Buckwheat flour-a valuable component of gluten-free formulations. *Polish journal of food and nutrition sciences, 58*(1).

Xie, S. X., Liu, Q., & Cui, S. W. (2005). *Starch modification and applications* (pp. 357–406). CRC Press, Boca Raton: FL.

Ziobro, R., Korus, J., Witczak, M., & Juszczak, L. (2012). Influence of modified starches on properties of gluten-free dough and bread. Part II: Quality and staling of gluten-free bread. *Food Hydrocolloids, 29*(1), 68–74.

# Chapter 4
# Role of Microbial Fermentation in Gluten-Free Products

**R. Anand Kumar and Winny Routray**

**Abstract** Gluten intolerance is one of the significant symptoms associated with different health disorders, which has become an increasing concern worldwide. A gluten-free diet is considered a curative product for the problem, which has been steadily increasing in the market. Gluten plays a key role in developing gluten-containing products with desired attributes. Elimination of gluten in staple food is not an easy task and it is difficult to provide the gluten-free product with similar characteristics as gluten-based products. Gluten-free products are produced from gluten-free cereals with different kinds of additives that modify the product according to the desired properties. Gluten-free cereals such as rice, sorghum, maize, and corn are some of the raw materials that are the major replacement cereals for producing gluten-free products. Fermentation is also an avital step in the preparation of the gluten-based product for attaining optimum texture and sensory properties. Sourdough fermentation is an important process employed in the fermentation of gluten-free products for creating resemblance with the gluten-based product. Lactic acid bacteria species have been mostly used in the fermentation process to produce gluten-free products of equivalent quality. Enzymes are also utilized in the production of non-gluten products such as gluten-free beer. The major drawbacks of gluten-free products include the cost of production, nutritional deficiency, and lack of simpler methods to produce the final products. Lack of nutrients in gluten-free products is due to the replacement of raw materials that can be recovered by the incorporation of nutrients from different sources such as vegetables and grains containing vitamins, minerals, and high-level dietary fiber. The modification of physical, chemical, and aromatic properties by the incorporation of additives also influences the final product quality. The future market scenario mainly depends on the adaptation of new lifestyle diets by the respective consumers for the respective abnormality and disease.

**Keywords** Gluten intolerance · Gluten-free · Sourdough fermentation · Lactic acid bacteria · Textural properties · Sensory and aromatic properties

---

R. Anand Kumar · W. Routray (✉)
Department of Food Process Engineering, NIT Rourkela, Rourkela, Odisha, India
e-mail: routrayw@nitrkl.ac.in



N. Singh Deora et al. (eds.), *Challenges and Potential Solutions in Gluten Free Product Development*, Food Engineering Series,
https://doi.org/10.1007/978-3-030-88697-4_4

## 4.1 Introduction

Gluten-free products, low-carb diets, and low-fat diets have become some of the most common and important terms used in the new dietary diaspora, where the trend of gluten-free diet consumption is increasing worldwide, attributed to the increased consciousness about health and dietary lifestyles. The market value of gluten-free products has been anticipated to reach 6.47 billion USD by 2023, which was 4.48 billion USD in 2018 (Markets, 2018). In the Indian market, it is expected to reach 189 million USD by 2024. It was 8.62 million USD in 2018 (Markets, 2019). Baked products are some of the most commonly consumed gluten-free products in the diet, wherein gluten-free coconut cookies and choco-chip cookies are some of the popular products in the Indian market and possibly throughout the world (Markets, 2019). Furthermore, e-commerce is one of the major players in the market of gluten-free products, which has given a common platform for the local and global manufacturers and augmented the selection of products for the consumers, combined impacting the market and the consumer wellbeing.

Gluten intolerance is one of the major concerns in recent days. Also, celiac disease is one of the commonly observed autoimmune diseases, mainly caused due to genetic disorders and environmental factors. It can be identified through gastrointestinal symptoms and extraintestinal manifestations (Torres et al., 2007). The disease is triggered mainly due to consuming food that contains gluten such as food made of wheat, barley, oat, and rye. Symptoms of celiac disease include bloating, vomiting, and diarrhea (Skerritt et al., 1990). There is also another condition called non-celiac gluten sensitivity, which incurs similar symptoms such as bloating, gas, abdominal pain, weight loss due to malabsorption, anemia, and fatigue (Murray et al., 2004; Ghadami, 2016). Celiac disease occurs due to particular genes, including HLA – DQ2 and DQ8 Haplotype. Allergens present in wheat protein have been characterized by a food IgE-mediated allergy. The important allergen and main toxic component present in wheat gluten are ω5-gliadin that induces anaphylaxis (Balakireva & Zamyatnin, 2016). The untreated celiac disease leads to osteoporosis, epilepsy and genetic problems such as Turner syndrome, Down syndrome and IgA deficiency (Ghadami, 2016).

Fermentation of food enhances the flavor and nutritional quality of food, where microorganisms play a vital role. Since, Louis Pasteur observed lactic acid fermentation in 1857 and 1837, and proposed that yeast is responsible for the conversion of sugars into ethanol and carbon dioxide, which are some of the main components contributing to the physicochemical properties of fermented products, the fermentation process has been diversified and has become an innate unit-operation for the processing of various food products. Egyptians, 4000 years ago, fermented the dough for the preparation of bread, which is currently available in different versions as a major staple food all over the world. Wheat bread, beer, pasta, cakes, cookies, and similar pastries are some of the fermented products containing gluten as one of the main raw materials. Furthermore, currently, different other fermented products are also available in the market. Sauerkraut is produced through the fermentation of cabbage, wherein the present method of sauerkraut fermentation was developed by Vaughn (1981).

The market value of the baking industry is expected to be 17 billion USD by2022, which was about 13 billion USD in 2017 worldwide. In India, the market value was 7.22 billion USD in 2018 (India 2020). India is the second-largest producer of biscuits and is expected to reach a market value of 12 billion USD in 2024. The market value of beer was 593 billion USD in 2017. It is expected to reach 685 billion USD in 2025. In India 13 billion USD is the current market value of beer (Statista, 2020). Hence, based on the popular consensus regarding baked and fermented products, it can be deduced that the demand for gluten-containing fermented and baked products will also further grow. However, attributed to the negative effects and health complications observed in cases of consumption of gluten-containing products, the development of gluten-free products is being continually encouraged.

This chapter has summarized different aspects and considerations for the development of gluten-free products, the microorganisms involved in the fermentation of gluten-free products, properties of these products, and the corresponding process and composition modifications required for obtaining consumer acceptable products, which have also been simultaneously briefly compared with the gluten-containing products. Prospective future perspective of the gluten-free products and challenges encountered for successful development of the acceptable products has also been discussed.

## 4.2 Different Currently Available Gluten-Free Products with Fermentation

### *4.2.1 Traditional and Non-traditional Products Derived from Gluten-Containing Grains*

Traditionally a wide array of gluten-containing products have been available, which are still some of the major sample subjects of studies. For the past many centuries, several baked products have been developed with different flavors. Gluten is the most important constituent in baked products, which can deform, stretch and trap air molecules inside the dough, useful for the production of bread and other similar fermented products. Bread is a major traditional product that is made of raw material containing gluten. Controlling bulk fermentation is the most important factor for the quality of bread (Cauvain, 2015). Cakes are batter-based and chemically leavened products and currently, there are several varieties of cakes, which have been developed through variation of formulation of the ingredients. The pasta varieties include macaroni, vermicelli, lasagna, spaghetti, and noodles. Semolina wheat is a major raw material for manufacturing pasta. Pizza, which is originally from Italy and is a flat leavened bread with different kinds of toppings, is produced from hard wheat flour, where dough develops a gluten network for entrapping carbon dioxide. Empanada is another food item, which can be combined with paneer and prepared using refined wheat flour, milk powder, shortening agent, yeast, and salt; this is also available inspired and sweet forms (Mallikarjuna, 2013). The tortilla is produced from wheat flour, shortening agent, salt, baking powder, and other essential additives with water, wherein the dough structure is maintained by gluten matrix. Gluten is a major factor for the development of a good quality product with desired texture and shelf stability (Alviola et al., 2008). Biscuits, cookies, and crackers are some of the other major products produced from refined wheat flour.

Though wheat-based gluten-containing products are the most commonly available products in the market other gluten-containing grains, from which several products have been developed include rye, barley, and triticale (a hybrid and rye). Barley has higher dietary fiber (2–11% β-glucan) than wheat (0.5–1%) (Feng et al., 2005) and has been recognized as a healthier alternative. Barley bread is made from 80 % of the whole meal barley flour and 20% of white wheat flour. These bread are consumed as part of breakfast by the healthy participants and have been assessed for glucose tolerance effect. It was found that insulin level was maintained at lower quantity for the participants (Östman et al., 2002). Rye bread is prepared in four varieties such as endosperm rye bread, whole-grain rye bread, whole-grain rye bread with lactic acid, and rye bran bread. These products have a significant effect on maintaining low blood glucose levels and stimulated low insulin (Rosén et al., 2009).

Apart from the regular popularly available gluten-based products, there are also several other products developed through fermentation. Fermentation of cracker dough has been observed to modify the protein network, wherein fermentation of dough is followed by cutting and sheeting (Zydenbos et al., 2004). In a different study, barley was used for enriching Tarhana with high β-glucan content, which is traditionally a fermented product prepared from the mixture of wheat flour and yogurt, and has originated from Turkey (Erkan et al., 2006). Awad and Salama (2010) added fermented barley in cheese produced from buffalo milk and buttermilk powder, to develop a product similar to the labneh-fermented (Greek Yoghurt) product.

There are also several traditional and non-traditional beverages derived from gluten-based grains. American rye whiskey is an important fermented alcoholic beverage product of rye grain. It has unique characteristics compared to other whiskeys, and are dry and spicy (Lahne, 2010). The raw spirit produced from the grains such as rye and wheat has been used for the manufacturing of the Vodka after rectification of the spirit (Lachenmeier et al., 2003). Benzoxazinoids is introduced in the barley-fermented beer by the addition of rye or wheat malt. It is a nitrogen-containing secondary metabolite compound, which has a positive effect on health including weight reduction, central nervous system stimulatory, antimicrobial, and immune-regulatory effects (Adhikari et al., 2015; Pihlava & Kurtelius, 2016).

### 4.2.2 Gluten-Free Products Developed from Gluten-Free Grains

Cereals not containing gluten include rice, maize, sorghum, and millets, which are also used for the production and formulation of gluten-free products. Correspondingly, celiac patients can consume gluten-free pasta made from buckwheat flour (Alamprese et al., 2007). Channa flour, soy flour, sorghum flour, and whey protein concentrate are the raw materials for producing gluten-free pasta (Susanna & Prabhasankar, 2013; Gao et al., 2018). These cereals have been considered as the major replacements of gluten-containing raw materials for producing gluten-free diets. In gluten-free biscuits, refined wheat flour is replaced by other raw materials, which should consist of not only starch but also the equal amount of protein fractions as present in the original recipe with wheat. Gluten-free biscuits produced from starch contain raw materials such as corn, rice, millet, potato, and buckwheat, mixed with fat like palm oil, low and high-fat dairy powder. The quality of biscuits is comparable to wheat flour biscuits when it is prepared with rice, soya, corn, and potato with high fat powder (Gallagher, 2008).

Different non-traditional food commodities have also been developed as healthy alternatives by several research groups, with enhanced sensory properties. Gluten-free empanada and pies were developed from cassava starch, dry egg, whey protein concentrate, gums, and water by Lorenzo et al. (2008). The addition of gum produced dough that had high elasticity properties and lesser hydration. These properties were reported as almost similar to the gluten-containing dough used in industries. Sorghum flour was used by Winger et al. (2014)for preparing gluten-free tortilla by hot press procedure, where sorghum flour completely replaced the wheat flour, and xanthan gum, baking powder, citric acid, sugar, monoglycerides, shortening agent, salt, and water were used for the preparation of tortillas. Rice-based muffins have also been prepared by adding different protein sources such as soy protein isolate, pea protein isolate, casein, egg white protein, which were compared with samples containing wheat gluten, for the assessment of samples win terms of conventional muffin properties. Muffins with pea protein isolate were found to be softer; whereas, casein containing muffin was harder. The texture quality of the muffin was greatly dependent on the protein sources (Matos et al., 2014).

Malting and brewing of gluten-free cereals such as sorghum, rice, and maize, is a common process for producing alcoholic beverages (Zweytick & Berghofer, 2009). Sake is an alcoholic beverage manufactured from rice and water. Sake is traditionally produced in Japan with steamed rice. The broken rice from the milling industry is also used for producing beer, wherein the sweet taste of rice beer changes to sour when sorghum is added to it; the taste is acceptable after successful organoleptic assessment (Phiarais & Arendt, 2008). There are several other low alcoholic beverages such as Braga, darassum, cochate (Phiarais & Arendt, 2008), and uphutsu, which is a traditional beer brewed using Pearl millet in Mozambique (Pelembe et al., 2002; Phiarais & Arendt, 2008).

### *4.2.3 Gluten-Free Products Derived from Gluten-Containing Grains*

The gluten content in bread is reduced by sourdough fermentation. Sourdough fermentation has been proven to be ideal for improving the shelf life and nutritional value of bread, where after mixing water with flour, the mixture is fermented with yeasts and lactic acid bacteria. The metabolic activities of the corresponding microorganisms (Lactobacilli) in the sourdough cause positive effects, including fermentation of lactic acid, proteolysis, production of exo-polysaccharides, and antimicrobial compounds synthesis (Corsetti & Settanni, 2007). Sourdough fermentation helps in reducing the glycemic index of the bread (Scazzina et al., 2009). Physio-chemical properties of the dough correspondingly modify, as sourdough fermentation also causes higher resistance to deformation after gelatinization of the batter (Schober et al., 2007).

Refined triticale flour has been used for bread making and other baked products. Triticale flour does not produce good quality dough due to low gluten content, high levels of alpha-amylase activity, and inferior gluten strength. These limitations have been overcome by mixing at low speed, shorter fermentation times, and blending with wheat flour to produce bread with acceptable quality by Naeem et al. (2002). Onwulata et al. (2000) also used triticale flour to produce extruded high fiber snacks and a nutritious bar containing 20–40% wheat and oat bran. However, oriental noodles manufactured from triticale flour possessed poorer properties, including the greyish color of noodles caused due to high ash content in flour creating an undesirable property (Shin et al., 1980).

Worldwide, there are several products developed from gluten-containing grains, which possess lower concentrations of gluten attributed to the fortification with other components. Miso is a Japanese product prepared by fermentation of barley and soybeans with *Aspergillus oryzae*, and the corresponding thick paste-like Miso product is a regular diet of the Japanese people. Routine consumption of miso has been reported to decrease the risk of gastric cancer (Murooka & Yamshita, 2008; Hirayama, 1982). However, due to the high level of salt, it possesses the risk of blood pressure (Kawano, 2007). A modified version of Tempeh (Indonesian soy product) is produced employing barley, wherein *Rhizopus oligosporous* and *Lactobacillus plantarum* are used for the fermentation process to produce barley tempeh (Feng et al., 2005).

Based on the above-mentioned sections, it can be observed that fermentation improved the quality of products, both with gluten and without gluten. Hence, it can be deduced that fermentation is an important and essential processing method for the development of gluten-free products, which are derived through reduction of gluten in products derived from gluten containing grains and products derived through formulation with gluten free grains.

## 4.3 Properties of the Products to Be Considered

The quality of products, including bread, biscuits, pastas, etc. mainly depends upon the raw material and processing methods. There is no individual characteristic that can be used for the sole identification of the quality of the product. There are many properties to be considered that decides the final quality of the product.

### *4.3.1 Compositions and Chemical Properties*

Traditional compositions and preferred gluten-containing products are generally affected by the component flour properties. Gluten network formation is the major process in bread making. The protein content of the flour determines the bread quality, where higher protein content enables to trap the carbon dioxide and retains the greater volume of bread. Dough properties such as specific volume, spread ratio, color, smoothness, texture depend on the protein quality (Zhu et al., 2001). Yeasts are available in different forms and the major type of yeast used in the baking is the baker's yeast (*Saccharomyces cerevisiae*) (Bell et al., 2001). Yeast feeds on sugar to release carbon dioxide (Ali et al., 2012) and produces carbon dioxide during the fermentation process for leavening of bread dough. Adding fats like margarine increases the carbon dioxide retention in the dough. Also, attributed to the formation of hydrogen peroxide by yeast, the dough becomes more elastic. However, insufficient hydration of dough leads to incomplete formation of gluten network (Faridi & Faubion, 2012); hence, textural properties of dough are modified through the amount of water added (Cauvain & Young, 2007). Salt controls the fermentation process by strengthening the network formed by gluten and increase of the effect of lipid peroxidation reaction (Toyosaki & Sakane, 2013).To improve the product quality, additives like oxidizing agents, reducing agents, emulsifiers and enzyme active materials are also added (Cauvain, 2015). Emulsifiers such as sodium stearoyl lactylate, lecithin and distilled monoglycerides are also used in bread making. Similarly, other products including biscuits and cookies are also produced with flour, sugar, fats and oil.

In gluten-free breads, soya flour is used to replace protein content and starch is added through rice and maize, which replace the essential amylopectin content (Taghdir et al., 2017). Eggs are often added to the gluten-free compositions as an emulsifier, which also contains proteins that enable strong cohesive viscoelastic films that are necessary substrates to trans glutaminase for the formation of stable foaming (Moore et al., 2006). Gluten-free cookies have also been made by using germinated amaranth flour, which has high antioxidant activity and total dietary fiber content (Chauhan et al., 2015).

Apart from regular compositions with other grains for starch and protein replacement, supplementary additives are also added with bioactive and textural properties. In a separate study, gluten-free rice muffins were developed using black carrot pomace dietary fiber concentrate and xanthan gum, which demonstrated higher water and oil absorption capacity and consequently increased flour paste viscosities (Singh et al., 2016).

Fermentation of gluten-free materials with microorganisms and sourdough fermentation in baked goods also enhances functional properties such as an increase in free amino-acid concentration, increase in antioxidant activity (Gobbetti et al., 2014; Curiel et al., 2015), and increase in mineral (free $Ca^{2+}$, $Zn^{2+}$and $Mg^{2+}$) availability (Di Cagno et al., 2008).

### *4.3.2 Physical Properties*

Based on the available reports and scientific papers, it can be concluded that some of the different desirable physical properties required for the bakeries and confectioneries and other related food products include textural properties such as hardness and fracturability values (Kadan et al., 2001), loaf volume and color (Sciarini et al., 2010) of the product.

During the past decade, different gluten-free grains have been used to prepare bread. In gluten-free bread, during fermentation, $CO_2$ is released due to the absence of the gluten network, which influences the specific volume, oven spring, and other characteristics of bread. This contributes to the gas retention properties and water absorption properties. Hence, the absence of gluten leads to bread with low volume and dense structure (Ayo, 2001). In bread made with rice flour (wherein amylopectin is a major component, which imitates the gluten properties), milled defatted bran, yeast, sugar, and salt, the hardness, and fracturability value are tenfold higher than the whole wheat bread. The values of properties, such as springiness, cohesiveness, and chewiness have been observed to reduce in whole rice bread during the storage period. Rice bread was not found to be acceptable for sandwich preparation by Kadan et al. (2001), as the rice bread is more brittle in nature. In a different study, Sciarini et al. (2010) prepared gluten-free bread with different formulations consisting of a combination of rice, corn, and soy, wherein the loaf volumes of the gluten-free bread were found to be lower than the conventional wheat bread. Furthermore, the addition of soy flour (10%) to the rice bread reduced the crumb hardness of the gluten-free bread, and the addition of cornflour darkened the color of the crust of the rice flour bread; however, the color was less dark than the normal wheat bread. Gluten-free donut have also been made with a combination of regular rice flour, pre-gelatinized rice flour, and reduced vanilla content in the ingredient mix. Xanthan gum and methylcellulose were added for replacing gluten. Methylcellulose had increased the L* and b* values in the crust color of the donut and a* value remains the same as the wheat donut (Melito & Farkas, 2013). In the case of cookies prepared with amaranth- oat composites significant water holding capacity was demonstrated (Inglett et al., 2015).

In pasta, strong gluten is necessary for maintaining products with less sticky properties and enhanced textural properties (Padalino et al., 2016). However, to achieve better sheeting properties, there is a need for weaker gluten and better extensible dough, as gluten matrix plays a vital role in providing desirable properties of the product. In gluten-free pasta, the desirable physicochemical properties are obtained by the application of additives, including hydrocolloids, proteins, and enzymes (Padalino et al., 2016). Similarly, in pizza, appearance, texture, and taste are important physical properties identified by a consumer, wherein the quality of pizza mainly depends on the characteristics of the dough, controlled by the leavening process. Gluten-free pizza is prepared with potato starch, wheat starch, maize starch, cornflour, rice flour, gums, and emulsifiers. Elastic properties of gluten-free dough were observed to increase with the amount of water and hydrocolloid in the flour mix (Onderi, 2013).

Other functional properties often considered during product development from gluten-containing and gluten products include emulsification, foaming, fat absorption, and thickening, which vary extensively with different compositions. Emulsification has been reported to decrease with an increase in soy protein content in the flour blends, which has been further reduced with pea protein content in the ingredient (Tömösközi et al., 2001).

Apart from the blends of different grains, the addition of different specific chemical additives and modifiers also affect the physical properties of the corresponding developed products, which include hydrocolloids, gums, and enzymes. Hydroxypropyl methylcellulose (HPMC) has been reported as one of the best replacements for gluten and it has been observed to enhance the properties of bread, including volume, moisture content, and decrease of the hardness of bread (Hager & Arendt, 2013). However, enzymes such as transglutaminase have been observed to decrease the foaming stability and foaming activity attributed to the increase in molecular activity as observed in the case of rice flour-based gluten-free product development (Marcoa & Rosell, 2008; Marco et al., 2007). Other additives and supplements have been discussed in detail in Sect. 4.6.

### 4.3.3 Aromatic Properties

The different aromatic components which contribute towards the traditional aromatic properties of the gluten-based products include ethyl acetate, ethyl butanoate, diacetyl, dihydrocoumarin, butyric acid, decanoic acid, benzaldehyde, vanillin, and propylene glycol, which are further modified with different additional additive components (Pozo-Bayón et al., 2006). The pyrazines and 2-acetyl-1-pyrroline are the most significant compound for producing the desirable aromatic properties of bread (Pacyński et al., 2015). 2-acetyl-1-pyrroline and methyl propanal are key odorants in the baguette crust (Cho & Peterson, 2010). In wheat bread crumb (E) – 2-Nominal is responsible for green tallow odor. $\gamma$ – Nonalactone is responsible for the coconut-like aroma and (E, Z)-2,6- non-adienal leads to cucumber-like aroma in bread. 2-Pentylfuran is formed during baking due to the process of fermentation, Maillard reactions, and lipid oxidation. 2-Pentylfuran is aromatic factor responsible for the floral fruity notes in the wheat bread crumb (Pico et al., 2017). In the rye bread crust, 3-methyl butanal is aroma producing factor.

Further addition of other gluten-free grains, such as oats, as an extra adjunct, provides modified flavor properties. In a study on assessment and improvement of the gluten-free bread aroma, about 33 volatile compounds detected in GC-MS from the gluten-free bread prepared using a bread mix made available by Glutenex (Pacyński et al., 2015). The volatile compounds identified included alcohols, aldehydes, ketones, pyrazines, and furans. However, pyrazines have been reported to be absent in gluten-free bread, which is traditionally found in wheat flour bread. Incorporation of precursors like proline in combination with glucose or fructose in gluten-free bread has been reported to enhance the aromatic properties by producing acetyl pyrazine. Furthermore, the presence of other components such as methional has also been identified in gluten-free bread. In a different study by Annan et al. (2003), during the production of kenkey (staple dish of West Africa and Ghana) from maize dough, about 76 aromatic compounds were identified within the period of the fermentation process. Esters like Ethyl acetate and Ethyl lactate were also produced in higher concentrations during the fermentation period of 2 days.

Hence, it can be observed that significant changes occur in the physiochemical and aromatic properties, during the production of gluten-free products; however, products with desirable properties can be obtained through the controlled formulation of food ingredients and additives.

## 4.4 Microbial Strains Useful for Gluten-Free Products

### *4.4.1 Traditional Microorganisms (Benefits and Characteristics)*

Gluten-containing fermented food items are some of the oldest developed and consumed grain-based food commodities; hence, a wide array of microorganisms is used for the synthesis of these. *Saccharomyces cerevisiae* is the most commonly used microorganism for the fermentation of bread dough and has also been employed in the production of wine. Lactic acid bacteria (LAB) has also been used for fermenting wheat and rye flour dough and traditional LAB strains reduce toxic and anti-nutritive factors in cereals (Holzapfel et al., 2006). The corresponding strains have been observed to remove stachyose, raffinose, and verbascose from soy-based products, which are the main cause of flatulence and intestinal cramps in case of higher oligosaccharides content (Dworkin et al., 2006; Holzapfel et al., 2006). *S.boulderi* and *pombe* are also some of the yeast used in the fermentation of traditional beverages, including white grape wine, cashew wine, red grape wine, banana beer, date wine, tepache, jackfruit wine, palm wine, and colonche (Battcock, 1998). The other fermenting bacterial species for producing several locally available exotic products from gluten-based cereals are *Leuconostoc*, *Pediococcus* and *Lactobacillus*

(Blandino et al., 2003), which have been often employed for the production of boza, dhokla, hamanatto, kecap, shoyu, and kenkey.

Fermentation also increases the bioavailability of mineral constituents and contributes towards the development of better organoleptic and aromatic properties. Application of the above-mentioned organisms has been mainly associated with the aromatic properties; however, other physiochemical properties are also affected by the fermentation methods and the chosen microorganism. Gamel et al. (2015) fermented with three different dough-making processes leading to the straight, sponge, and sourdough. The bread prepared from straight and sponge dough had higher molecular weight and decreased β glucan content compared to sourdough, which was attributed to the fermentation of dough with microorganisms such as yeast and lactic acid bacteria in sourdough bread leading to lower pH (4.2–4.6), wherein high acidity in dough maintained the molecular weight of β-glucan and decreased the activity of degrading enzymes of β-glucan (Bhatty, 1992). However, an increase in fermentation time decreased the molecular weight of β-glucan and viscosity in the bread, as increased fermentation time leads to greater contact time with degrading enzymes. Due to enzyme degradation of amylose and amylopectin during fermentation of wheat starch, a reduction in molecular weight of wheat and amylose has also been observed (Nowak et al., 2014).

Fermentation has also been associated with an increase in the content of bioactive compounds such as phenolics, as often observed in the cases of alcoholic beverages. Fermentation with *Cornus Officinalis* has been observed to increase phenolic compounds from 441.06 to 496.00 mg/L in wines. An increase in the amount of gallic acid observed during the fermentation process has been correlated with metabolites produced by yeast (Zhang et al., 2013).

### *4.4.2 Microorganisms Utilized in Gluten-Free Products*

*Lactobacillus plantarum* has been used for producing gluten-free bread, consisting of buckwheat flour, cornstarch, soy flour, and xanthan gum through the method of sourdough fermentation. *L. Plantarum*also produced antifungal compounds that enabled shelf life extension of the bread (Moore et al., 2008). It produces acetic acid from pentose sugar and lactic acid from hexoses. In sorghum flour pentose sugar content is very low as compared to hexoses; hence, lactic acid production is higher during the sorghum flour fermentation (Schober et al., 2007). Instant dry yeast has also been used for the preparation of gluten-free bread from rice and maize flour. *Aspergillus oryzae* and yeast have been used to produce gluten-free bread from the rice flour by Hamada et al. (2013), where the batter was fermented at 38 °C for 60 min.

In a different study, *Lactobacillus reuteri* and *Weissella cibaria*were used for the production of fermented sorghum and quinoa flours for further food applications. The fermentation process produced fructo-oligosaccharides and gluco-oligosaccharides. *Lactobacillus buchneri* released hetero polysaccharides, which significantly controlled the rheology properties of sorghum sourdough, according to a study by Galle et al. (2011). In a separate study, *Leuconostoc argentinum, Pediococcus pentosaceus,* and *Weisella cibaria* were the species identified for the fermentation of oat sourdough for preparing oat bread, wherein *Lactobacillus coryniformis* was also identified and shown to cause a higher reaction rate during the fermentation process at 37 °C (Hüttner et al., 2010). In the case of sorghum fermented for the towga production, which is a traditional Tanzanian lactic acid fermented gruel, lactic acid bacteria are the predominant microorganism, which also leads to the production of various flavor compounds (Zannini et al., 2012). Select lactic acid bacteria can also be used for enriching the gluten-free products with γ-Aminobutyric acid, which is a major non-protein amino-acid acting as an "inhibitory neurotransmitter of the central nervous system" (Coda et al., 2010).In a separate study by Di Cagno et al. (2008), *Lactobacillus plantarum* and *Lactobacillus sanfranciscensis* LS40 and LS41were used for the fermentation process of gluten-free flour, where the gluten content decreased from 400 ppm to 20 ppm.

Hence, though lactic acid bacteria have been extensively exploited for the development of gluten-free products, several other bacterial strains have also been identified and employed for gluten-free product development.

## 4.5 Processing Methods and Apparatus Employed in Case of Gluten-Free Product Development

Sourdough technology is a traditional method used for the production of wheat and rye breads, wherein overall the process consisted of mixing flour with water followed by fermentation using lactic acid bacteria and yeast, causing the release of lactic and acetic acid and producing sour-tasting end product (Chavan & Chavan, 2011). Lactic acid bacteria are applied in desired proportions for enhancing volume, texture, nutritional value, and flavor along with the enhancement of the shelf life of the bread. Although the standard method of sourdough production is a simple method, researchers have modified and optimized the process parameters for achieving different targets in the case of the production of gluten-free products. During a study by Moore et al. (2008) on the development of sourdough preparation method from gluten-free flour mixture, cultures of *Lacto bacillus* were inoculated at 1% concentration levels in 40 ml of broth, which was incubated for 1 day at 30 °C and was centrifuged (4000 rpm for 5 min) to prepare concentrated harvested culture. This extracted strain was again dispersed in 1 ml of broth and mixed with water followed by mixing with flour, after which, the well-mixed batter was incubated at 30 °C for about 24 h.

Most of the processing methods and apparatus used in gluten-free product development are similar, the part that varies most is replacing the raw material, and inclusion of new additives for the development of the product that is equivalent to the gluten-based product. Extrusion processing has been used for enriching gluten-free products with high-level dietary fiber. Processing conditions of the extruder considered by Stojceska et al. (2010) included feed rate of 15–25 kg/h, barrel temperatures at hopper side as 80 °C and 80–150 °C at die point along with screw speed between 200 and 350 rpm. During this study, the product from the die output was cut using a knife attached at the end of the die, wherein the product was later cooled and left at room temperature, and subsequently packed. The process was similar to the processing of pasta, spaghetti, and other similar products obtained through the extrusion process, with enhanced nutritional value through the addition of different sources of raw materials (Stojceska et al., 2010; Zhang et al., 2011). Apparatus used for this process was a co-rotating twin-screw extruder with a barrel diameter of 37 mm and an L/D ratio of 27:1. The diameter of the die was 4 mm and a volumetric feeder was used for feeding the dry mixture (Stojceska et al., 2010).

Germination and enzymatic treatment of the respective grains and their corresponding extracts have also been employed for the enhancement of nutritional value and decrease of gluten concentrations, which are also useful processes for producing gluten-free products. In a different study by Knorr et al. (2016), barley was germinated using a proofing cabinet during the batch process, after which the steeping process was carried out for producing gluten-free beer through peptidase treatment. After germination, malt was extracted from sprout, which was followed by the milling process. Milled malt was extracted with water using a magnetic stirring process, which was subsequently filtered and the filtered extracts were concentrated using a rotary evaporator set at 50 °C and 65 mbar. The final product was obtained at the concentration of 40° brix, wherein, the final enzyme active malt extract was used for the production of beer. Beer was also produced using barley malt and hop pellets by fermenting for 7 days using yeast at 10 °C, where secondary fermentation processes were carried for 4 weeks at 4 °C after the addition of wort, which was incubated with enzyme active malt extract (10% (v/v)) at 50 °C for 1 day (Knorr et al., 2016). Enzyme treatment in the production of gluten-free beer is essential. Enzyme hydrolyses the peptide linkages (peptide sequences are responsible for producing the celiac disease) that occur during downstream processing of proline residue. Beers have also been produced from gluten-free cereals such as sorghum, corn, millet, and rice (Hager et al., 2014).

## 4.6 Additives and Modifiers Used for Gluten-Free Products

### *4.6.1 Different Additives and Modifiers*

Different kinds of additives, such as emulsifiers, enzymes, and hydrocolloids have been used for the development of gluten-free products (including breads), which are also useful for enhancing the quality of the final product. Hydrocolloids have been increasingly used in bread making process. These compounds act as additives to slow down the retrogradation of starch and are also used for the replacement off at, for the enhancement of viscoelastic properties and texture. These components bind water molecules and contribute towards the maintenance of the quality of the product during storage. The quality of bread produced from rye flour was enhanced through the addition of guar gum and carboxymethyl cellulose (CMC) by Ghodke Shalini and Laxmi (2007). Other components extensively employed in product development include the emulsifiers such as diacetyl tartaric acid ester of monoglyceride and sodium stearoyl 2-lactylate, enzymes such as glucose oxidase and α – amylase, and other hydrocolloids like xanthan gum, carboxymethyl cellulose, carrageenan and alginate (Sciarini et al., 2012).

Microorganisms are mainly added for fermentation during product development; however, in many cases, they are also added for fortification. LAB has demonstrated antifungal activity and these microorganisms have been reported to prevent the staling of bread; hence, application of these organisms can be proposed as a better alternative for chemical preservatives (Lowe & Arendt, 2004). Specifically targeted extracts can also be prepared, which demonstrate beneficial properties. The Amaranthus seeds extract contain peptides that have been reported to demonstrate antifungal activity, wherein it was observed that the substrate significantly retarded the growth of *Penicillium roqueforti* for 21 days of storage (Giuseppe Rizzello et al., 2009); hence, the addition of these extracts in the composition of gluten-free product formulation can also be further extensively exploited.

### 4.6.2 Modification of the Chemical and Biological Properties of the Gluten-Free Products Attributed to the Additives

Chemical and biological additives modify the physio-chemical properties of the raw, intermediary and final products, affect the storage life as well consumer acceptability of the final products. Furthermore, the extent of the effect is also affected by the combination of the effects of individual components and effects of other processing and surrounding factors, where the final product properties are the resultant of the amalgamation of all the different factors.

Chemical properties of the products and their overall makeup, significantly affected by the additives, include the chemical and physical structures/ transformations, intermediate conversions, bonding and stability of the components. HPMC and starch hydrolyzing enzymes have been observed to reduce amylopectin retrogradation (Bárcenas & Rosell, 2007) in the corresponding compositions such as rice-based breads (Gujral et al., 2003a, b).Transglutaminase, which is another helpful additive, has the capability of linking proteins originated from different sources. Transglutaminase has been used in gluten free bread by Moore et al. (2006), which consisted of soy, egg and skim milk powder, wherein transglutaminase helps in forming stable protein network without the gluten compound. The protein network helps in developing the good quality gluten free bread with desirable loaf volume, crumb characteristics and texture.

Stability at different stages of processing and storage are also significantly affected by the components and their proportions. Xanthan gum and pectin have been reported to improve cooking stability (Kohajdová et al., 2009),whereas, calcium propionate used for preventing fungal growth at time of bread storage (Tosh et al., 2012). Inulin is a prebiotic compound digested by colon bacteria, which at 5% concentration with oligosaccharide syrup and chicory flour has been observed to decrease the staling rate for 3 days, as reported by Korus et al. (2006). It also retarded rate of crumb hardening and subsequently improved the loaf volume. Enzymes like α – amylase and cyclodextrin glycosyl transferase extracted from *Bacillus* species reduced staling rate of rice bread (Gujral et al., 2003a, b). Similarly, the components produced by the microorganisms during the process of fermentation not only affect the physiochemical and sensory properties relevant for the product quality in terms of consumer acceptance, but also the storage quality of the developed product through the production of disparate biochemicals. Bacteriocin produced by lactic acid bacteria has been observed to delay the growth of *Aspergillus flavus* and *Aspergillus niger* in the gluten free cereal product called *agidi* (Dike & Sanni, 2010).

The different additives also contribute towards further enhancing the health beneficial effects of these gluten-free products. Inulin has been observed to enhance the dietary fiber content in the gluten free foods, which subsequently enhances the digestive health of the consumer (Korus et al., 2006). Also, in other cases, essential health beneficial components are enriched through fortification of the additives. Folate content is very low in gluten free product compared to gluten containing product and it is essential to maintain the concentrations of folate and other vitamins in the diet for maintaining good health. Hence, for the development of gluten free products, pseudo-cereals like quinoa and amaranth species can be included, where these components can provide ten fold higher concentrations of folate as compared to other cereals such as spring wheat (Schoenlechner et al., 2010). There is need of vitamin and mineral source to be added in gluten free product, as it provides less concentrations of these nutrients as compared to conventional diet.

### 4.6.3 Modification of Physical Properties of the Gluten-Free Products Attributed to the Additives

Physical properties of the developed gluten-free products are significantly affected by the absence of gluten, as discussed in the previous sections. However, consumer acceptability of these products can be significantly improved through the addition of modifiers improving the physical properties of these products.

Stability of dough is an essential qualitative entity for the indication of flour strength. Incorporation of hydrocolloids like carrageenan, xanthan gum and hydroxy-propyl methylcellulose (HPMC) has been observed to improve the stability of wheat dough at the time of proofing (Sahraiyan et al., 2013). Hence, firmness of breadcrumb has been observed to reduce through the addition of κ-carrageenan or HPMC. Addition of HPMC has also led to the reduced hardening of the bread, enhanced specific volume index and width/height ratio. In a different study, addition of 2% xanthan gum reduced porosity of the bread that provided good textural quality of bread (Lazaridou et al., 2007). Crumb elasticity of the gluten free bread improved by adjunct of CMC, pectin and xanthan gum that improved textural properties (Arendt et al., 2008). Furthermore, Schober et al. (2005) observed increased loaf volume and crumb hardness for the bread produced from sorghum with addition of xanthan gum, wherein increase in crumb hardness was as an undesirable quality of the bread, which demonstrated the occurrence of negative effects of factors as well in certain cases. Addition of hydrocolloid in bread with 65% water also led to inferior properties such as dense crumb structure, low specific volume and high firmness, which could be recovered by the addition of higher water content (Sciarini et al., 2012). Hence, though there are negative effects of absence of gluten and certain additives, through optimized combination of different components, the desirable properties of the commodities can be retrieved and/or achieved.

### 4.6.4 Modification and Enhancement of Aromatic Properties

Modification of aroma and flavor greatly depends on the raw material, type of starter cultures, fermentation and baking conditions. Change in raw material from wheat to other similar ingredients leads to lack of aromatic compounds such as pyrazines and 2-acetyl-1-pyrroline, which were originally present in gluten based products (Zehentbauer & Grosch, 1998). These limitations of aroma can be overcome by manual addition of different kinds of amino acids and sugar pairs to the raw materials before processing. In order to produce aroma, pairs of precursor compounds for targeted aroma were added by Pacyński et al. (2015), including proline/glucose, leucine/glucose, proline/fructose, ornithine/fructose and cysteine/rhamnose. Also, incorporation of high level of sodium chloride (1.5–3%) has been associated with decrease in the amount of 2-phenylethanol, a compound causing smell of yeast (Raffo et al., 2018). Addition of different grains and their corresponding compositional proportions can also lead to different aromatic profiles of the products. In a study by Wolter et al. (2014), where buckwheat, quinoa, sorghum, teff and wheat breads, quinoa and sorghum flour (and teff crumb) were observed to produce "cooked potato and pea" like and "cooked tomato and pea" like odors, respectively. Overall, the different additives and their corresponding concentrations used for producing aroma is based on the consumer likeability and preference.

Microorganisms employed during the different unit-operations of the product development have also been associated with development of various aromatic compounds. Increase in amount of yeast during the preparation of the product has been observed to enhance the production of 2-acetyl-1-pyrroline and methional. These compounds have been identified as the main factors responsible for causing roast like aroma (Zehentbauer & Grosch, 1998). In a different study by Wolter et al. (2014), *Weissella cibaria*has been observed to produce the 'popcorn –like roasty' aroma in gluten-free breads . Quinoa flour was observed to produce 'cooked potato and pea' like and 'cooked tomato and pea like odour produced for sorghum and teff crumb (Wolter et al., 2014).

## 4.7 Future Perspectives

### *4.7.1 Combined Benefits of Fermentation and Gluten-Free Compositions*

Gluten free diet has less fiber compared to the normal diet. Hence, food technologists and dieticians have been increasingly recommending the consumption of gluten free baked products with the addition of dietary fibers. The ancient grains like amaranth seeds have higher beneficial effects that satisfy the requirement for wheat replacement (Fornal, 2000); however, there is a need for the optimization of the fermentation process for newer as well as non-traditional raw materials. Established methods such as fermentation and extrusion are being further optimized to obtain high quality gluten-free products. Fermentation increases digestibility and absorption of nutrients, which is also one of the low-cost preservation methods (Beyene and Seifu). The process also detoxifies the products and provides better textural properties. Gluten free composition in fermented products provides the combined effect of fermentation benefit and gluten free raw materials used in the product (Chojnacka, 2010).

Subsequently, these benefits are further increased through the addition of other novel and non-traditional components; studies on these topics have also increased recently. The incorporation of bee pollen in gluten free bread has been observed to increase total carotenoids, proteins, minerals, soluble and bio-accessible polyphenols and antiradical activity at all levels (2–5%) (Conte et al., 2020). Most of these bioactive components have added benefits, which include both primary and secondary metabolic functions. Carotenoids are the precursor of vitamin A and provide protection against most of eye-related chronic diseases (Rodriguez-Concepcion et al., 2018; Johnson, 2002). Dietary polyphenols are essential for the maintenance of gut health and balance of gut microbiota. It enhances the growth of beneficial bacteria and prohibits pathogenic bacteria (Cardona et al., 2013). They also provide other beneficial effects including antioxidant activity and effectiveness against several other chronic disorders such as diabetes, different cancers, neurodegenerative and ocular disorders (Orsat & Routray, 2017).

### 4.7.2 *Possible Future Market Scenario*

Gluten free diets can be used for the treatment of medical conditions such as bowel syndrome, arthritis, dermatitis, diabetes mellitus and other neurological disorders (Wahnschaffe et al., 2007; Badsha, 2018; Sanchez-Albisua et al., 2005). Many consumers choose gluten free diet for their healthier life style. Gluten free products have been recognized as high quality products. An increase in number of consumers for gluten-free diets has led to novel products and corresponding process development to achieve the various qualities of the products based on consumer demand and acceptability. Hence, it can be deduced that gluten-free diets have to be designed and prepared, according to the different medical conditions observed in people along with consumer demand. For example, for diabetic consumer, gluten-free diet with a low glycemic index should be developed. In accordance with the medical conditions, the ingredients and the processing techniques employed in the manufacturing process will be different for obtaining the final desired product. It is expected that in the future, gluten free cereal-based probiotic products will achieve huge growth in market and these products will target consumers with either or both the gluten and lactose intolerance.

## 4.8 Conclusions

Attributed to the higher growth rate in industrialization and urbanization, there is a large demand for processed foods. Food consciousness in consumers has increased which has also led to higher consumption of gluten free food products. Gluten concentration influences a wide range of properties subsequently affecting the overall final product quality. There is a need for the supplementation of multiple ingredients to replace the single wheat flour and to avoid the gluten introduction in the product manufacturing process. Hydrocolloids are important additives required for the production of good quality gluten-free products. Both, starch-based and protein-based ingredient are necessary for the formulation of gluten-free products; hence, the addition of the old grains and a wide array of other raw materials for the replacement of wheat flour is essential, wherein the old grains and other dietary additives will have higher nutritional value and satisfy the dietary needs. Most of the lactic acid bacteria generate bacteriocins that can be used for the replacement of chemical preservatives. Also, new microorganism strains have to be produced with help of genetic engineering for achieving further innovation in this sector and further optimization of the parameters of the fermentation process is also necessary to achieve high-quality gluten-free food products. Further research is also required to analyze the consequences of consuming a gluten-free diet containing additives such as xantham gum, HPMC, and other similar components that help in providing good quality gluten-free diet.

## References


Adhikari, K. B., Tanwir, F., Gregersen, P. L., Steffensen, S. K., Jensen, B. M., Poulsen, L. K., Nielsen, C. H., Høyer, S., Borre, M., & Fomsgaard, I. S. (2015). Benzoxazinoids: Cereal phytochemicals with putative therapeutic and health-protecting properties. *Molecular Nutrition & Food Research, 59*(7), 1324–1338. https://doi.org/10.1002/mnfr.201400717

Alamprese, C., Casiraghi, E., & Pagani, M. A. (2007). Development of gluten-free fresh egg pasta analogues containing buckwheat. *European Food Research and Technology, 225*(2), 205–213. https://doi.org/10.1007/s00217-006-0405-y

Ali, A., Shehzad, A., Khan, M. R., Shabbir, M. A., & Amjid, M. R. (2012). Yeast, its types and role in fermentation during bread making process-A. *Pakistan Journal of Food Sciences, 22*(3), 171–179.

Alviola, J. N., Waniska, R. D., & Rooney, L. W. (2008). Role of gluten in flour tortilla staling. *Cereal Chemistry, 85*(3), 295–300. https://doi.org/10.1094/CCHEM-85-3-0295

Annan, N. T., Poll, L., Plahar, W. A., & Jakobsen, M. (2003). Aroma characteristics of spontaneously fermented Ghanaian maize dough for kenkey. *European Food Research and Technology, 217*(1), 53–60. https://doi.org/10.1007/s00217-003-0697-0

Arendt, E. K., Morrissey, A., Moore, M. M., & Bello, F. D. (2008). 13 – Gluten-free breads. In E. K. Arendt & F. Dal Bello (Eds.), *Gluten-free cereal products and beverages* (p. 289). Academic. https://doi.org/10.1016/B978-012373739-7.50015-0

Awad, R. A., & Salama, W. M. (2010). Development of a novel processed cheese product containing fermented barley. *Egyptian Journal of Dairy Science, 38*(1), 95–103.

Ayo, J. A. (2001). The effect of amaranth grain flour on the quality of bread. *International Journal of Food Properties, 4*(2), 341–351. https://doi.org/10.1081/JFP-100105198

Badsha, H. (2018). Role of diet in influencing rheumatoid arthritis disease activity. *The Open Rheumatology Journal, 12*, 19.

Balakireva, A. V., & Zamyatnin, A. A. (2016). Properties of gluten intolerance: Gluten structure, evolution, pathogenicity and detoxification capabilities. *Nutrients, 8*(10), 644.

Bárcenas, M. E., & Rosell, C. M. (2007). Different approaches for increasing the shelf life of partially baked bread: Low temperatures and hydrocolloid addition. *Food Chemistry, 100*(4), 1594–1601. https://doi.org/10.1016/j.foodchem.2005.12.043

Battcock, M. (1998). *Fermented fruits and vegetables: A global perspective*. Food & Agriculture Organisation.

Bell, P. J. L., Higgins, V. J., & Attfield, P. V. (2001). Comparison of fermentative capacities of industrial baking and wild-type yeasts of the species Saccharomyces cerevisiae in different sugar media. *Letters in Applied Microbiology, 32*(4), 224–229. https://doi.org/10.1046/j.1472-765X.2001.00894.x

Bhatty, R. S. (1992). Total and extractable β-glucan contents of oats and their relationship to viscosity. *Journal of Cereal Science, 15*(2), 185–192. https://doi.org/10.1016/S0733-5210(09)80070-2

Blandino, A., Al-Aseeri, M. E., Pandiella, S. S., Cantero, D., & Webb, C. (2003). Cereal-based fermented foods and beverages. *Food Research International, 36*(6), 527–543. https://doi.org/10.1016/S0963-9969(03)00009-7

Cardona, F., Andrés-Lacueva, C., Tulipani, S., Tinahones, F. J., & Queipo-Ortuño, M. I. (2013). Benefits of polyphenols on gut microbiota and implications in human health. *The Journal of Nutritional Biochemistry, 24*(8), 1415–1422. https://doi.org/10.1016/j.jnutbio.2013.05.001

Cauvain, S. (2015). Breadmaking processes. In *Technology of breadmaking* (pp. 23–55). Springer.

Cauvain, S. P., & Young, L. S. (2007). Technology of breadmaking.

Chauhan, A., Saxena, D. C., & Singh, S. (2015). Total dietary fibre and antioxidant activity of gluten free cookies made from raw and germinated amaranth (Amaranthus spp.) flour. *LWT – Food Science and Technology, 63*(2), 939–945. https://doi.org/10.1016/j.lwt.2015.03.115

Chavan, R. S., & Chavan, S. R. (2011). Sourdough technology – A traditional way for wholesome foods: A review. *Comprehensive Reviews in Food Science and Food Safety, 10*(3), 169–182. https://doi.org/10.1111/j.1541-4337.2011.00148.x

Cho, I. H., & Peterson, D. G. (2010). Chemistry of bread aroma: A review. *Food Science and Biotechnology, 19*(3), 575–582. https://doi.org/10.1007/s10068-010-0081-3

Chojnacka, K. (2010). Fermentation products. *Chemical Engineering and Chemical Process Technology*, 12

Coda, R., Rizzello, C. G., & Gobbetti, M. (2010). Use of sourdough fermentation and pseudo-cereals and leguminous flours for the making of a functional bread enriched of γ-aminobutyric acid (GABA). *International Journal of Food Microbiology, 137*(2), 236–245. https://doi.org/10.1016/j.ijfoodmicro.2009.12.010

Conte, P., Del Caro, A., Urgeghe, P. P., Petretto, G. L., Montanari, L., Piga, A., & Fadda, C. (2020). Nutritional and aroma improvement of gluten-free bread: Is bee pollen effective? *LWT, 118*, 108711. https://doi.org/10.1016/j.lwt.2019.108711

Corsetti, A., & Settanni, L. (2007). Lactobacilli in sourdough fermentation. *Food Research International, 40*(5), 539–558. https://doi.org/10.1016/j.foodres.2006.11.001

Curiel, J. A., Coda, R., Centomani, I., Summo, C., Gobbetti, M., & Rizzello, C. G. (2015). Exploitation of the nutritional and functional characteristics of traditional Italian legumes: The potential of sourdough fermentation. *International Journal of Food Microbiology, 196*, 51–61. https://doi.org/10.1016/j.ijfoodmicro.2014.11.032

Di Cagno, R., Rizzello, C. G., De Angelis, M., Cassone, A., Giuliani, G., Benedusi, A., Limitone, A., Surico, R. F., & Gobbetti, M. (2008). Use of selected sourdough strains of Lactobacillus for removing gluten and enhancing the nutritional properties of gluten-free bread. *Journal of Food Protection, 71*(7), 1491–1495.

Dike, K. S., & Sanni, A. I. (2010). Influence of starter culture of lactic acid bacteria on the shelf life of agidi, an indigenous fermented cereal product. *African Journal of Biotechnology, 9*(46), 7922–7927.

Dworkin, M., Falkow, S., Rosenberg, E., Schleifer, K. H., & Stackebrandt, E. (2006). *The genera pediococcus and tetragenococcus. The prokaryotes* (pp. 229–266). Springer, US Press.

Erkan, H., Çelik, S., Bilgi, B., & Köksel, H. (2006). A new approach for the utilization of barley in food products: Barley tarhana. *Food Chemistry, 97*(1), 12–18. https://doi.org/10.1016/j.foodchem.2005.03.018

Faridi, H., & Faubion, J. M. (2012). *Dough rheology and baked product texture*. Springer Science & Business Media.

Feng, X. M., Eriksson, A. R. B., & Schnürer, J. (2005). Growth of lactic acid bacteria and Rhizopus oligosporus during barley tempeh fermentation. *International Journal of Food Microbiology, 104*(3), 249–256. https://doi.org/10.1016/j.ijfoodmicro.2005.03.005

Fornal, J. (2000). Structural properties of starch in food systems. *Żywność, 2*(23), 59–71.

Gallagher, E. (2008). 14 – Formulation and nutritional aspects of gluten-free cereal products and infant foods. In E. K. Arendt & F. Dal Bello (Eds.), *Gluten-free cereal products and beverages* (pp. 321–346). Academic. https://doi.org/10.1016/B978-012373739-7.50016-2

Galle, S., Schwab, C., Arendt, E. K., & Gänzle, M. G. (2011). Structural and rheological characterisation of heteropolysaccharides produced by lactic acid bacteria in wheat and sorghum sourdough. *Food Microbiology, 28*(3), 547–553. https://doi.org/10.1016/j.fm.2010.11.006

Gamel, T. H., Abdel-Aal, E.-S. M., & Tosh, S. M. (2015). Effect of yeast-fermented and sour-dough making processes on physicochemical characteristics of β-glucan in whole wheat/oat bread. *LWT – Food Science and Technology, 60*(1), 78–85. https://doi.org/10.1016/j.lwt.2014.07.030

Gao, Y., Janes, M. E., Chaiya, B., Brennan, M. A., Brennan, C. S., & Prinyawiwatkul, W. (2018). Gluten-free bakery and pasta products: Prevalence and quality improvement. *International Journal of Food Science & Technology, 53*(1), 19–32. https://doi.org/10.1111/ijfs.13505

Ghadami, M. R. (2016). Celiac disease and epilepsy: The effect of gluten-free diet on seizure control. *Pediatrics, 3*, 52–08.

Ghodke Shalini, K., & Laxmi, A. (2007). Influence of additives on rheological characteristics of whole-wheat dough and quality of Chapatti (Indian unleavened Flat bread) Part I – hydrocolloids. *Food Hydrocolloids, 21*(1), 110–117. https://doi.org/10.1016/j.foodhyd.2006.03.002

Giuseppe Rizzello, C., Coda, R., De Angelis, M., Di Cagno, R., Carnevali, P., & Gobbetti, M. (2009). Long-term fungal inhibitory activity of water-soluble extract from Amaranthus spp. seeds during storage of gluten-free and wheat flour breads. *International Journal of Food Microbiology, 131*(2-3), 189–196. https://doi.org/10.1016/j.ijfoodmicro.2009.02.025

Gobbetti, M., Rizzello, C. G., Di Cagno, R., & De Angelis, M. (2014). How the sourdough may affect the functional features of leavened baked goods. *Food Microbiology, 37*, 30–40. https://doi.org/10.1016/j.fm.2013.04.012

Gujral, H. S., Guardiola, I., Carbonell, J. V., & Rosell, C. M. (2003a). Effect of cyclodextrinase on dough rheology and bread quality from rice flour. *Journal of Agricultural and Food Chemistry, 51*(13), 3814–3818. https://doi.org/10.1021/jf034112w

Gujral, H. S., Haros, M., & Rosell, C. M. (2003b). Starch hydrolyzing enzymes for retarding the staling of rice bread. *Cereal Chemistry, 80*(6), 750–754. https://doi.org/10.1094/CCHEM.2003.80.6.750

Hager, A.-S., & Arendt, E. K. (2013). Influence of hydroxypropylmethylcellulose (HPMC), xanthan gum and their combination on loaf specific volume, crumb hardness and crumb grain characteristics of gluten-free breads based on rice, maize, teff and buckwheat. *Food Hydrocolloids, 32*(1), 195–203. https://doi.org/10.1016/j.foodhyd.2012.12.021

Hager, A.-S., Taylor, J. P., Waters, D. M., & Arendt, E. K. (2014). Gluten free beer – A review. *Trends in Food Science & Technology, 36*(1), 44–54. https://doi.org/10.1016/j.tifs.2014.01.001

Hamada, S., Suzuki, K., Aoki, N., & Suzuki, Y. (2013). Improvements in the qualities of gluten-free bread after using a protease obtained from Aspergillus oryzae. *Journal of Cereal Science, 57*(1), 91–97. https://doi.org/10.1016/j.jcs.2012.10.008

Hirayama, T. (1982). Relationship of soybean paste soup intake to gastric cancer risk. *Nutrition and Cancer, 3*(4), 223–233. https://doi.org/10.1080/01635588109513726

Holzapfel, W. H., Franz, C., Ludwig, W., Back, W., & Dicks, L. M. (2006). The genera pediococcus and tetragenococcus. *Prokaryotes, 4*, 229–266.

Hüttner, E. K., Dal Bello, F., & Arendt, E. K. (2010). Identification of lactic acid bacteria isolated from oat sourdoughs and investigation into their potential for the improvement of oat bread quality. *European Food Research and Technology, 230*(6), 849–857. https://doi.org/10.1007/s00217-010-1236-4

Inglett, G. E., Chen, D., & Liu, S. X. (2015). Physical properties of gluten-free sugar cookies made from amaranth–oat composites. *LWT – Food Science and Technology, 63*(1), 214–220. https://doi.org/10.1016/j.lwt.2015.03.056

Johnson, E. J. (2002). The role of carotenoids in human health. *Nutrition in Clinical Care, 5*(2), 56–65. https://doi.org/10.1046/j.1523-5408.2002.00004.x

Kadan, R. S., Robinson, M. G., Thibodeaux, D. P., & Pepperman, A. B., Jr. (2001). Texture and other physicochemical properties of whole rice bread. *Journal of Food Science, 66*(7), 940–944. https://doi.org/10.1111/j.1365-2621.2001.tb08216.x

Kawano, K. (2007). History and functional components of miso. *Nippon Aji Nioi Gakkaishi, 14*, 137–144.

Knorr, V., Wieser, H., & Koehler, P. (2016). Production of gluten-free beer by peptidase treatment. *European Food Research and Technology, 242*(7), 1129–1140. https://doi.org/10.1007/s00217-015-2617-5

Kohajdová, Z., Karovičová, J., & Schmidt, Š. (2009). Significance of emulsifiers and hydrocolloids in bakery industry. *Acta Chimica Slovaca, 2*(1), 46–61.

Korus, J., Grzelak, K., Achremowicz, K., & Sabat, R. (2006). Influence of prebiotic additions on the quality of gluten-free bread and on the content of inulin and fructooligosaccharides. *Food Science and Technology International, 12*(6), 489–495. https://doi.org/10.1177/1082013206073072

Lachenmeier, D. W., Attig, R., Frank, W., & Athanasakis, C. (2003). The use of ion chromatography to detect adulteration of vodka and rum. *European Food Research and Technology, 218*(1), 105–110. https://doi.org/10.1007/s00217-003-0799-8

Lahne, J. (2010). Aroma characterization of American rye whiskey by chemical and sensory assays.

Lazaridou, A., Duta, D., Papageorgiou, M., Belc, N., & Biliaderis, C. G. (2007). Effects of hydrocolloids on dough rheology and bread quality parameters in gluten-free formulations. *Journal of Food Engineering, 79*(3), 1033–1047. https://doi.org/10.1016/j.jfoodeng.2006.03.032

Lorenzo, G., Zaritzky, N., & Califano, A. (2008). Optimization of non-fermented gluten-free dough composition based on rheological behavior for industrial production of "empanadas" and piecrusts. *Journal of Cereal Science, 48*(1), 224–231. https://doi.org/10.1016/j.jcs.2007.09.003

Lowe, D. P., & Arendt, E. K. (2004). The use and effects of lactic acid bacteria in malting and brewing with their relationships to antifungal activity, mycotoxins and gushing: A review. *Journal of the Institute of Brewing, 110*(3), 163–180. https://doi.org/10.1002/j.2050-0416.2004.tb00199.x

Mallikarjuna, B. M. (2013). Technological studies on the development of paneer empanada.

Marco, C., Pérez, G., Ribotta, P., & Rosell, C. M. (2007). Effect of microbial transglutaminase on the protein fractions of rice, pea and their blends. *Journal of the Science of Food and Agriculture, 87*(14), 2576–2582. https://doi.org/10.1002/jsfa.3006

Marcoa, C., & Rosell, C. M. (2008). Effect of different protein isolates and transglutaminase on rice flour properties. *Journal of Food Engineering, 84*(1), 132–139. https://doi.org/10.1016/j.jfoodeng.2007.05.003

Markets Ma. (2018). *Gluten-free products market.* https://www.marketsandmarkets.com/Market-Reports/gluten-free-products-market-738.html#:~:text=The%20gluten%2Dfree%20products%20market%20is%20estimated%20to%20be%20valued,and%20food%20intolerances%20among%20consumers

Markets Ra. (2019). India gluten-free foods & beverages market – Growth, trends and forecast (2019–2024). *Research and Markets.* https://www.researchandmarkets.com/reports/4602381/india-gluten-free-foods-and-beverages-market?utm_source=dynamic&utm_medium=BW&utm_code=ktk5m3&utm_campaign=1336375+-+India+Gluten-Free+Foods+%26+Beverages+Market+Report+2019-2024&utm_exec=chdo54bwd

Matos, M. E., Sanz, T., & Rosell, C. M. (2014). Establishing the function of proteins on the rheological and quality properties of rice based gluten free muffins. *Food Hydrocolloids, 35*, 150–158. https://doi.org/10.1016/j.foodhyd.2013.05.007

Melito, H., & Farkas, B. E. (2013). Physical properties of gluten-free donuts. *Journal of Food Quality, 36*(1), 32–40. https://doi.org/10.1111/jfq.12008

Moore, M. M., Heinbockel, M., Dockery, P., Ulmer, H. M., & Arendt, E. K. (2006). Network formation in gluten-free bread with application of transglutaminase. *Cereal Chemistry, 83*(1), 28–36. https://doi.org/10.1094/CC-83-0028

Moore, M. M., Bello, F. D., & Arendt, E. K. (2008). Sourdough fermented by Lactobacillus plantarum FST 1.7 improves the quality and shelf life of gluten-free bread. *European Food Research and Technology, 226*(6), 1309–1316. https://doi.org/10.1007/s00217-007-0659-z

Murooka, Y., & Yamshita, M. (2008). Traditional healthful fermented products of Japan. *Journal of Industrial Microbiology & Biotechnology, 35*(8), 791. https://doi.org/10.1007/s10295-008-0362-5

Murray, J. A., Watson, T., Clearman, B., & Mitros, F. (2004). Effect of a gluten-free diet on gastrointestinal symptoms in celiac disease. *The American Journal of Clinical Nutrition, 79*(4), 669–673.

Naeem, H. A., Darvey, N. L., Gras, P. W., & MacRitchie, F. (2002). Mixing properties, baking potential, and functionality changes in storage proteins during dough development of triticale-wheat flour blends. *Cereal Chemistry, 79*(3), 332–339. https://doi.org/10.1094/CCHEM.2002.79.3.332

Nowak, E., Krzeminska-Fiedorowicz, L., Khachatryan, G., & Fiedorowicz, M. (2014). Comparison of molecular structure and selected physicochemical properties of spelt wheat and common wheat starches. *Journal of Food & Nutrition Research, 53*(1).

Onderi, M. O. (2013). Effects of xanthan gum and added protein on the physical properties of gluten-free pizza dough: A texture characterization study using instron model. 3342.

Onwulata, C. I., Konstance, R. P., Strange, E. D., Smith, P. W., & Holsinger, V. H. (2000). High-fiber snacks extruded from triticale and wheat formulations. *Cereal Foods World, 45*(10), 470–473.

Orsat V, Routray W (2017) Microwave-assisted extraction of flavonoids. In: H. D. Gonzalez & M. J. G. Munoz (Eds.), *Water extraction of bioactive compounds from plants to drug development* ( 1st ed., pp. 221–244). Elsevier.

Östman, E. M., Liljeberg Elmståhl, H. G. M., & BjÖrck, I. M. E. (2002). Barley bread containing lactic acid improves glucose tolerance at a subsequent meal in healthy men and women. *The Journal of Nutrition, 132*(6), 1173–1175. https://doi.org/10.1093/jn/132.6.1173

Pacyński, M., Wojtasiak, R. Z., & Mildner-Szkudlarz, S. (2015). Improving the aroma of gluten-free bread. *LWT – Food Science and Technology, 63*(1), 706–713. https://doi.org/10.1016/j.lwt.2015.03.032

Padalino, L., Conte, A., & Del Nobile, M. A. (2016). Overview on the general approaches to improve gluten-free pasta and bread. *Food, 5*(4), 87.

Pelembe, L. A. M., Dewar, J., & Taylor, J. R. N. (2002). Effect of malting conditions on pearl millet malt quality. *Journal of the Institute of Brewing, 108*(1), 7–12. https://doi.org/10.1002/j.2050-0416.2002.tb00113.x

Phiarais, B. P. N., & Arendt, E. K. (2008). 15 – Malting and brewing with gluten-free cereals. In E. K. Arendt & F. Dal Bello (Eds.), *Gluten-free cereal products and beverages* (pp. 347–372). Academic. https://doi.org/10.1016/B978-012373739-7.50017-4

Pico, J., Martínez, M. M., Bernal, J., & Gómez, M. (2017). Evolution of volatile compounds in gluten-free bread: From dough to crumb. *Food Chemistry, 227*, 179–186. https://doi.org/10.1016/j.foodchem.2017.01.098

Pihlava, J.-M., & Kurtelius, T. (2016). Determination of benzoxazinoids in wheat and rye beers by HPLC-DAD and UPLC-QTOF MS. *Food Chemistry, 204*, 400–408. https://doi.org/10.1016/j.foodchem.2016.02.148

Pozo-Bayón, M. A., Guichard, E., & Cayot, N. (2006). Feasibility and application of solvent assisted flavour evaporation and standard addition method to quantify the aroma compounds in flavoured baked matrices. *Food Chemistry, 99*(2), 416–423. https://doi.org/10.1016/j.foodchem.2005.08.005

Raffo, A., Carcea, M., Moneta, E., Narducci, V., Nicoli, S., Peparaio, M., Sinesio, F., & Turfani, V. (2018). Influence of different levels of sodium chloride and of a reduced-sodium salt substitute on volatiles formation and sensory quality of wheat bread. *Journal of Cereal Science, 79*, 518–526. https://doi.org/10.1016/j.jcs.2017.12.013

Rodriguez-Concepcion, M., Avalos, J., Bonet, M. L., Boronat, A., Gomez-Gomez, L., Hornero-Mendez, D., Limon, M. C., Meléndez-Martínez, A. J., Olmedilla-Alonso, B., Palou, A., Ribot, J., Rodrigo, M. J., Zacarias, L., & Zhu, C. (2018). A global perspective on carotenoids: Metabolism, biotechnology, and benefits for nutrition and health. *Progress in Lipid Research, 70*, 62–93. https://doi.org/10.1016/j.plipres.2018.04.004

Rosén, L. A. H., Silva, L. O. B., Andersson, U. K., Holm, C., Östman, E. M., & Björck, I. M. E. (2009). Endosperm and whole grain rye breads are characterized by low post-prandial insulin response and a beneficial blood glucose profile. *Nutrition Journal, 8*(1), 42. https://doi.org/10.1186/1475-2891-8-42

Sahraiyan, B., Naghipour, F., Karimi, M., & Davoodi, M. G. (2013). Evaluation of Lepidium sativum seed and guar gum to improve dough rheology and quality parameters in composite rice–wheat bread. *Food Hydrocolloids, 30*(2), 698–703. https://doi.org/10.1016/j.foodhyd.2012.08.013

Sanchez-Albisua, I., Wolf, J., Neu, A., Geiger, H., Wäscher, I., & Stern, M. (2005). Coeliac disease in children with Type 1 diabetes mellitus: The effect of the gluten-free diet. *Diabetic Medicine, 22*(8), 1079–1082. https://doi.org/10.1111/j.1464-5491.2005.01609.x

Scazzina, F., Del Rio, D., Pellegrini, N., & Brighenti, F. (2009). Sourdough bread: Starch digestibility and postprandial glycemic response. *Journal of Cereal Science, 49*(3), 419–421. https://doi.org/10.1016/j.jcs.2008.12.008

Schober, T. J., Messerschmidt, M., Bean, S. R., Park, S.-H., & Arendt, E. K. (2005). Gluten-free bread from sorghum: Quality differences among hybrids. *Cereal Chemistry, 82*(4), 394–404. https://doi.org/10.1094/CC-82-0394

Schober, T. J., Bean, S. R., & Boyle, D. L. (2007). Gluten-free sorghum bread improved by sourdough fermentation: Biochemical, rheological, and microstructural background. *Journal of Agricultural and Food Chemistry, 55*(13), 5137–5146. https://doi.org/10.1021/jf0704155

Schoenlechner, R., Wendner, M., Siebenhandl-Ehn, S., & Berghofer, E. (2010). Pseudocereals as alternative sources for high folate content in staple foods. *Journal of Cereal Science, 52*(3), 475–479. https://doi.org/10.1016/j.jcs.2010.08.001

Sciarini, L. S., Ribotta, P. D., León, A. E., & Pérez, G. T. (2010). Influence of gluten-free flours and their mixtures on batter properties and bread quality. *Food and Bioprocess Technology, 3*(4), 577–585. https://doi.org/10.1007/s11947-008-0098-2

Sciarini, L. S., Ribotta, P. D., León, A. E., & Pérez, G. T. (2012). Incorporation of several additives into gluten free breads: Effect on dough properties and bread quality. *Journal of Food Engineering, 111*(4), 590–597. https://doi.org/10.1016/j.jfoodeng.2012.03.011

Shin, H.-K., Bae, S.-H., & Park, M.-Y. (1980). Nutritional quality and food-making performance of some triticale lines grown in Korea. *Korean Journal of Food Science and Technology, 12*(1), 59–65.

Singh, J. P., Kaur, A., & Singh, N. (2016). Development of eggless gluten-free rice muffins utilizing black carrot dietary fibre concentrate and xanthan gum. *Journal of Food Science and Technology, 53*(2), 1269–1278. https://doi.org/10.1007/s13197-015-2103-x

Skerritt, J. H., Devery, J. M., & Hill, A. S. (1990). Gluten intolerance: chemistry, celiac-toxicity, and detection of prolamins in foods. *Cereal Foods World, 35*(7), 638–644.

Statista. (2020). https://www.statista.com/outlook/10010000/119/beer/india

Stojceska, V., Ainsworth, P., Plunkett, A., & İbanoğlu, Ş. (2010). The advantage of using extrusion processing for increasing dietary fibre level in gluten-free products. *Food Chemistry, 121*(1), 156–164. https://doi.org/10.1016/j.foodchem.2009.12.024

Susanna, S., & Prabhasankar, P. (2013). A study on development of Gluten free pasta and its biochemical and immunological validation. *LWT – Food Science and Technology, 50*(2), 613–621. https://doi.org/10.1016/j.lwt.2012.07.040

Taghdir, M., Mazloomi, S. M., Honar, N., Sepandi, M., Ashourpour, M., & Salehi, M. (2017). Effect of soy flour on nutritional, physicochemical, and sensory characteristics of gluten-free bread. *Food Science & Nutrition, 5*(3), 439–445. https://doi.org/10.1002/fsn3.411

Tömösközi, S., Lásztity, R., Haraszi, R., & Baticz, O. (2001). Isolation and study of the functional properties of pea proteins. *Food / Nahrung, 45*(6), 399–401. https://doi.org/10.1002/1521-3803(20011001)45:6<399::AID-FOOD399>3.0.CO;2-0

Torres, M. I., López Casado, M. A., & Ríos, A. (2007). New aspects in celiac disease. *World Journal of Gastroenterology, 13*(8), 1156–1161. https://doi.org/10.3748/wjg.v13.i8.1156

Tosh, S. M., Ahmadi, L., Yip, L., Roudsari, M., & Wood, P. J. (2012). Presence of β-glucanase activity in wheat and dairy ingredients and use of organic salts as potential enzyme inhibitors. *Journal of Cereal Science, 56*(3), 538–543. https://doi.org/10.1016/j.jcs.2012.08.013

Toyosaki, T., & Sakane, Y. (2013). Effects of salt on wheat flour dough fermentation. *Advance Journal of Food Science and Technology, 5*(2), 84–89.

Vaughn, R. H. (1981). *Lactic acid fermentation of cabbage, cucumbers, olives and other products. Prescott and Dunn's industrial microbiology* (pp. 220–224). Saybrook Press.

Wahnschaffe, U., Schulzke, J. D., Zeitz, M., & Ullrich, R. (2007). Predictors of clinical response to gluten-free diet in patients diagnosed with diarrhea-predominant irritable bowel syndrome. *Clinical Gastroenterology and Hepatology, 5*(7), 844–850. https://doi.org/10.1016/j.cgh.2007.03.021

Winger, M., Khouryieh, H., Aramouni, F., & Herald, T. (2014). Sorghum flour characterization and evaluation in gluten-free flour tortilla. *Journal of Food Quality, 37*(2), 95–106. https://doi.org/10.1111/jfq.12080

Wolter, A., Hager, A.-S., Zannini, E., Czerny, M., & Arendt, E. K. (2014). Influence of dextran-producing Weissella cibaria on baking properties and sensory profile of gluten-free and wheat breads. *International Journal of Food Microbiology, 172*, 83–91. https://doi.org/10.1016/j.ijfoodmicro.2013.11.015

Zannini, E., Pontonio, E., Waters, D. M., & Arendt, E. K. (2012). Applications of microbial fermentations for production of gluten-free products and perspectives. *Applied Microbiology and Biotechnology, 93*(2), 473–485. https://doi.org/10.1007/s00253-011-3707-3

Zehentbauer, G., & Grosch, W. (1998). Crust aroma of baguettes II. Dependence of the concentrations of key odorants on yeast level and dough processing. *Journal of Cereal Science, 28*(1), 93–96. https://doi.org/10.1006/jcrs.1998.0183

Zhang, M., Bai, X., & Zhang, Z. (2011). Extrusion process improves the functionality of soluble dietary fiber in oat bran. *Journal of Cereal Science, 54*(1), 98–103. https://doi.org/10.1016/j.jcs.2011.04.001

Zhang, Q.-A., Fan, X.-H., Zhao, W.-Q., Wang, X.-Y., & Liu, H.-Z. (2013). Evolution of some physicochemical properties in Cornus officinalis wine during fermentation and storage. *European Food Research and Technology, 237*(5), 711–719. https://doi.org/10.1007/s00217-013-2045-3

Zhu, J., Huang, S., Khan, K., & O'Brien, L. (2001). Relationship of protein quantity, quality and dough properties with chinese steamed bread quality. *Journal of Cereal Science, 33*(2), 205–212. https://doi.org/10.1006/jcrs.2000.0358

Zweytick, G., & Berghofer, E. (2009). 10 production of gluten-free beer. *Gluten-Free Food Science and Technology, 181*(1).

Zydenbos, S., Humphrey-Taylor, V., & Wrigley, C. (2004). *Cookies, biscuits, and crackers| The diversity of products.*

# Chapter 5
# Functionality of Alternative Proteins in Gluten Free Product Development: Case Study

**Mahipal Singh Tomar, Sumit Sudhir Pathak, and Rama Chandra Pradhan**

**Abstract** Coeliac disease (CD), dermatitis herpetiformis (DH), non-coeliac gluten sensitivity (NCGS), and gluten ataxia (GA) are some of the most important problem, auto-immuno, and lifelong intolerance disorders in human. These disorders are found by the ingestion of gluten in our body. Gluten is mostly present in wheat, barley, rye, and other related grains. Replacement of gluten in our diet is the best method to reduce the chances the coeliac disease. Gluten replacement presents a major technological challenge, as it is an important protein that creates the structure required to formulate to bake the food of high quality. The functionality of non-gluten protein is the major limitation in the development of gluten free products. Finding of alternative protein is great demand in gluten free food markets. The selection of appropriate protein for gluten-free product is a great challenge for food industry. The current chapter focuses on the uses of alternative proteins to replace gluten. As well as studies related to the functionality and nutritional qualities of these alternative proteins are also discussed.

**Keywords** Coeliac disease · Gluten-free · Zein · Casein · Whey protein · Chickpea · Bread

## 5.1 Introduction

Coeliac disease (CD), dermatitis herpetiformis (DH), non-coeliac gluten sensitivity (NCGS), and gluten ataxia (GA) are some of the important, auto-immuno and lifelong intolerance disorders found by the ingestion of gluten in our body. It is mostly seen in genetically susceptible or peoples suffering from gluten

M. S. Tomar · S. S. Pathak · R. C. Pradhan (✉)
Department of Food Process Engineering, National Institute of Technology Rourkela, Rourkela, Odisha, India
e-mail: pradhanrc@nitrkl.ac.in



N. Singh Deora et al. (eds.), *Challenges and Potential Solutions in Gluten Free Product Development*, Food Engineering Series,
https://doi.org/10.1007/978-3-030-88697-4_5

intolerance. The Prevelencing of CD continuously increasing and almost it affected 0.5–1% of the worldwide population (Gujral et al., 2012; Deora et al., 2014). This gluten intolerance problem leads to the destruction of the villous structure of the small intestine and instigates the inflammatory problem. Gluten is a proteinous mixture of wheat (commonly consumed food), rye, barley cereals, and some varieties of oats. Gluten is made of different prolamins fraction, which is rich in prolamins and glutamine amino acids. Higher proline content in gluten, make it extremely resistant to proteolytic degradation inside of the gastrointestinal tract. Gliadin is another amino acid in gluten that is insoluble in water. And it is responsible for the most adverse health effects since it is a toxics factor for CD patients (Wieser, 1996). To reduce the effect of gluten on peoples and their adverse health effect, need a requirement of several high-quality non-gluten products. Hence researchers are more focusing on Gluten-Free product development. Presently, the consumption of gluten-free diet is alone one treatment for CD (Deora et al., 2015; Jerome et al., 2019). Functionality of non-gluten protein is the major limitation in development of non-gluten products. It has been great demand for finding alternative protein sources for the replacement of gluten and which should exhibit similar or more functional and nutritional properties to gluten. The current chapter focuses on the uses of alternative proteins to replace the gluten. As well as study related to the functionality and nutritional qualities of these alternative proteins are also discussed.

## 5.2 Protein Sources Other Than Gluten

### *5.2.1 Cereals*

Recent research support that idea of the functionalizing of non-wheat cereal proteins to imitate the viscoelastic nature of gluten is a promising field in the area of the development of gluten-free products (Deora et al., 2014). Cereal technologists around the world have successfully solved the problem of elimination of gluten from bread, biscuits, and other bakery products (Pradhan et al., 2021). Among all gluten-free products, bread is the most complex and commonly used baked product. The reason behind of complexity of bread is due to the role of gluten protein in the development of the bread matrix (Jerome et al., 2019). Rice flour is the most commonly used cereal flour for gluten-free bread making. It is also most suitable for various gluten-free products i.e. pasta, muffins, etc. due to its white color, easily digestible, and low prolamins content make it suitable for the patient suffering from coeliac disease. Various properties of products like the texture, appearance, and volume are improved and created by the use of different additive i.e. hydrocolloids, enzyme, protein acids, and emulsifiers. Storage protein of cereal is a good alternative for gluten replacement. Amino acid composition of zein and kafirin contains less percentage of prolamins and glutamine and it has unique properties like gluten for gluten-free product formulation. And it has been a great demanded research area for gluten-free product development.

### 5.2.2 Zein

Zein is the storage and functional properties of maize. Maize protein consists of 8–11% of whole weight, and zein is one of the major proteins found in the maize kernel. Zein protein is highly soluble in alcohol i.e. ethanol, ketones, glycerol solution but insolvable in water. Based on solubility and structural differences, zein protein is fractionated in four subclasses (Esen, 1987): α, and β-zein are major protein, γ and δ-zein are minor protein (Deora et al., 2015). α–zein is the highest (71–85% of total protein) protein composition of all zein subclasses (Lending & Larkins, 1989). Due to the presence of a large number of hydrophobic amino acid such as alanine, leucine, phenylalanine, and proline, α–zein significantly shows the hydrophobic properties (Gianazza et al., 1977).

Zein proteins have the ability for development of wheat (gluten) like dough and which has been successfully explored to produce gluten-free bread for CD sufferer's patient (Andersson et al., 2011; Jeong et al., 2017). At room temperature, zein protein is not able to form viscoelastic nature like wheat gluten, but higher temperature made it functional. Glass transition temperature of zein protein plays a vital role in the formation of viscoelastic nature. α–zein has the capability to the formation of viscoelastic properties, if the temperature of mixing and tempering was 35 °C and which is above the glass transition temperature (28 °C at ≥20 moisture content ) of zein. Viscoelastic material formation capability of zein protein can suggest because of non-covalent interaction between low molecular weight protein, and in the case of gluten is due to physical and chemical interaction of very high molecular protein (Smith et al., 2014). Viscoelastic properties of zein can also be modified by certain additive or modification to the protein themselves. This improvement in zein protein is possible either in alone protein or with the addition of co-protein.(Andersson et al., 2011) revealed that zein protein alone could not imitate the similar character of gluten dough but adding of hydrocolloids significantly improves the rheological and structural properties of zein dough. Zein protein dough shows a similar function to a wheat dough. By measurement of hyperbolic flow contraction of zein-starch dough with hydrocolloids shows the high extension viscosity. And this dough is best suitable for gluten-free bread development. During the baking process of dough, hydrocolloids supplemented zein-starch bread evolved the improved bread height, volume, and fine structure.

Development of dough by zein-rice starch mixture with amylose content is a possible alternative for the gluten-free product (Jeong et al., 2017). High amylose content in zein-rice paste exhibited great pasting and elasticity. And this mixture was suitable for the preparation of gluten-free noodles. Prepared noodles showed a firm texture.

### 5.2.3 *Kafirin*

Kafirin is the storage protein of sorghum and it shows great potential in the research area of alternative protein sources for gluten-free products. Sorghum endosperm contains 77–82% kafirin of total protein (De Mesa-Stonestreet et al., 2010). Kafirins are classified as prolamins and they also contain a high amount of glutamine and proline. It is insoluble in water and highly solvable in alcohol. The hydrophobic capacity of kafirin is higher among average protein and wheat's prolamins (Deora et al., 2015). Oom et al. (2008) found that kafirin protein has the capability for the formation of viscoelastic dough as wheat gluten. Strain hardening and extensive viscosity of kafirin protein were similar to wheat flour dough. They also revealed that kafirin dough exhibited adequate rheological properties for the development of gluten-free porous bread. Heat treatment of sorghum flour also affects achieving a higher volume of gluten-free cake and bread during the baking process (Marston et al., 2016).

## 5.3 Dairy and Poultry

In the recent year, applications of milk and egg protein have been increased in gluten-free food markets. Both ingredients can be used for alternative protein source in non- gluten product development. Incorporation of these alternative proteins improved the nutritional quality and structure of the non-gluten product. Casein, whey protein, and egg albumin protein provide the strength to gluten-free products.

### 5.3.1 *Whey Protein*

Whey protein concentrate (WPC) are extensively used dairy protein in research field of non-gluten products. The percentage of whey protein is 20% of total protein in milk (De Wit, 1990). Essential amino acid score of whey protein is higher than egg and soy protein. It is also considered as natural food additives for thickening functionality and as an alternative thickeners of starch and hydrocolloids (Resch & Daubert, 2002). Rheological properties of dough can be modify by addition of WPC in to dough paste (Lupano, 2003). van Riemsdijk et al. (2011) used the whey protein particle of meso-structured as a substitute of gluten for bread development. This structure of protein was selected based on Gluten free bread prepared by whey protein showed the strain hardening structure. Crumb structure of developed bread was similar to wheat bread (van Riemsdijk et al., 2011). Author also concludes that formulation of more or another protein with additives can improve the properties of the non-gluten bread. Volume of whey powder formulated bread was higher than the vital gluten based bread. Even small of whey amount (2.5%) was sufficient for large volume bread. Whey protein isolate enhanced the emulsifying properties, gelling behavior, and gelatinization of rice starch (Marcoa & Rosell, 2008). Ungureanu-Iuga et al. (2020) developed the gluten free pasta with whey powder and grape peel. Incorporation of whey powder in pasta paste improved the texture and sensory characteristics of pasta. Whey protein also enhanced the microstructure of pasta.

### *5.3.2 Casein*

Casein is a major milk protein and its percentage is highest approximately 80% of total milk protein. Casein is obtained by precipitation of milk at 4.6 pH, supernatant portion is called whey and another substance of precipitated milk is casein protein (Liang & Luo, 2020). It is commonly used as a binding agent for many food product developments. It comes under the group of phosphoproteins (Deora et al., 2015). Most important function of casein protein is used as an ingredient for structure building agents. They can be used in solid and semisolid food material to provide mechanical strength and improved textural properties. Proteins are also used as a thickening agent to boost the consistency and stability of developed food products (Chan et al., 2007; Deora et al., 2015). Caseinate is another important casein-based ingredient. Based on the emulsifying and foaming functionality of caseinate, it can be used to improve the properties and functionality of dough and paste for gluten free products (Luo et al., 2014). A combination of sodium caseinate and whey protein improved the quality of gluten free pasta of pearl millet (Kumar et al., 2019).

## 5.4 Egg Protein

Eggs are common food additives for food product development. In recent years, its application is increasing for gluten free product developments. Egg proteins are highly functional protein, and can be incorporate for gluten free dough formation (Ziobro et al., 2013; Crockett et al., 2011; Pico et al., 2019). Egg white protein is good alternative source for gluten. Based on functional properties i.e. foaming characteristics, improvement in crumb structure can replace the gluten from bakery products. For example, egg protein helps in the improvement of dispersion capability and stabilization of gas bubbles in the non-gluten dough system (Deora et al., 2015).

Egg albumin and whey protein with parboiled rice were used for the development of non-gluten pasta (Marti et al., 2014). The addition of protein significantly reduced the roughness of uncooked pasta. These proteins can be used as texturing agents in gluten free pasta production without addition of chemical agents. During pasta development, hydrophobic interaction and disulfide bond between rice starch and protein make the pasta stable. During the cooking of pasta, the formation of a disulfide interprotein bond takes place and result in significantly enhanced textural and structural properties of products (Marti et al., 2014).

Matos et al. (2014) developed the muffins with various protein sources. The emulsifying activity of rice flour was effectively increased by the addition of egg albumin. The best appearance of muffins was observed in the case of egg albumin and casein protein. The incorporation of egg white protein increases the height and volume of muffins. The author also found that animal-source protein produced chewy, springy, and more cohesive muffins than vegetable protein.

## 5.5 Legumes

Legumes are the plant seed in the family of Leguminosae. It is one of the most prominent sources of food protein. Primarily, legumes are grown for human consumption and but also for livestock feeding. Legumes are a good alternative and supplement grain for cereals based food products. Due to the presence of higher content of essential amino acid i.e. arginine, aspartic acid, lysine, glutamic acid, in legume, it can provide well sufficient diet with consumption of other cereals (Deora et al., 2015; Miñarro et al., 2012). Leguminous protein also has functional properties that play major role in the formation and processing of food (Boye et al., 2010). Legume protein has been used in development of various food product i.e. bakery product, soup, and several ready to eat snacks. Nutritional benefits of legume consumption suggest it's for alternative of gluten flour to the development of non-gluten food products (Miñarro et al., 2012).

Addition of legume flour or protein explored as a substitute for nutritional quality improvement as well as physical attributes and overall qualities of gluten free products (Crockett et al., 2011; Miñarro et al., 2012; Foschia et al., 2017; Pico et al., 2019). The functional properties of soy protein and pea protein isolate were used to make and developed various gluten free products (Deora et al., 2015).

## 5.6 Soya Protein

Soybean is one the most commonly grown and used oilseed. It is also rich source of protein and its percentage approximately 40% of total weight (Sharma et al., 2014). Also of a higher content of protein, the nutritional value of this protein is high. Its protein digestibility-corrected amino acid score is also in the region of egg white protein (Deora et al., 2015). Due to foam stabilization capacity of soya protein isolates, its incorporation in gluten free products are studied (Marcoa & Rosell, 2008; Crockett et al., 2011; Miñarro et al., 2012).

Rice flour-based muffins were developed by (Matos et al., 2014) with the addition of various gluten free protein sources for modification of properties of non-gluten muffins. The addition of soya protein isolates significantly increases the storage modulus of developed batter. It also altered the textural properties of baked products. In addition, its incorporation also modified viscosity and elastic component of the rice based batter, and results in inducing the hardening effect. In study of (Crockett et al., 2011), addition of soya protein increased the crumb structure and volume of bread. Soya protein and hydroxypropyl methylcellulose as an ingredient for rice flour bread development produced the bread with similar porosity like wheat gluten (Srikanlaya et al., 2018).

## 5.7 Carob Seed Germ Protein (Caroubin)

Caroubin is separated from the carob germ powder. Carob is evergreen flowing tree in the legume family of Fabaceae. Carob tree commonly cultivated for its edible pot. Germ of carob seed is rich source of protein and it's known as Caroubin. It is a mixture of great amount of protein number which varies in their size and degree of polymerization, ranging from one million to several thousands of molecular weight. Physico-chemical properties of this protein are thoroughly similar to gluten protein of wheat (Feillet & Roulland, 1998). Caroubin is not similar to wheat gluten but its function behaves like in the manner of wheat gluten (Smith et al., 2012). Due to the presence of disulfide bond between the high molecular proteins, caroubin germ flour produced a similar dough like wheat flour (Smith et al., 2012). And also the rheological properties of developed dough were similar to the gluten dough. The concentration of carob powder and water give significant viscoelastic characteristics of dough (Tsatsaragkou et al., 2014). Appropriate ratio of both parameters can produce dough with equilibrium viscosity and elasticity.

## 5.8 Chickpea Protein

Chickpea is one of the most consumable and important legume crops in Indian states. India is the highest production county of chickpea around the whole world. Chickpea plays a leading role in world food safety by resolving the problem of deficiency of protein in our daily diet (Kaur & Singh, 2007; Merga & Haji, 2019). Chickpea is a common and good source of protein and carbohydrate, and quality of protein is better than other legume crops such as pigeon pea, green gram, and black gram. Chickpea contain almost 40% protein of total mass. Due to such amount of protein, make it unique legume for food consumption. Besides protein content, it also has potential health benefits for reducing the cardiovascular, cancer, and diabetic risks (Kaur & Singh, 2005). Chickpea protein has various functional protein such as emulsifying and foaming characteristics (Boye et al., 2010). Due to functional properties, good nutrional qualities, and excellent baking capabilities of chickpea protein, it is good alternative protein source for gluten free products (Boye et al., 2010; Kaur & Singh, 2007; Aguilar et al., 2015). Addition of chick pea flour significantly increases the specific volume as well as storage modules of bread (Aguilar et al., 2015). The incorporation of chickpea flour with tiger nut flour can replaced the emulsifier and shortening agents for non-gluten-free product development.

## 5.9 Pseudocereals

The dicotyledonous plants which resemble the true cereals in their functions and composition are called pseudocereals. These pseudocereals can be classified as legumes, oilseeds, cereals and nuts, etc. Now a day's these pseudocereals are grabbing the attention of researchers to use them in gluten free product formulations since they have many health promoting effects as well as they possess lot of proteins, fibers, calcium and iron. If these pseudocereals are used for the formulation of gluten free product, there will no need to fortify the product by external addition of the minerals. Hence the usage of pseudocereals will make the product gluten free and will improve its nutritional quality (Alvarez-Jubete et al., 2010; Aghamirzaei et al., 2013).

Due to the high nutritional value and absolute free from the toxins while it possesses a considerable amount of protein. Majorly the amaranthus, quinoa and buckwheat are pseudocereals used for the formulation of gluten free products (Ballabio et al., 2011). In the studies conducted by many researchers, they have concluded that buck wheat and quinoa possesses high quality of protein which has good digestibility, balanced efficiency ratio. The protein from these sources resembles the qualities similar to milk proteins (Ranhotra et al., 1993; Repo-Carrasco et al., 2003). Some studies also showed that amaranth and quinoa are rich in bioactive compounds such as γ- and β-tocopherol as well as polyunsaturated fatty acids - high linolenic: linoleic acid ratio (Comino et al., 2013).

### *5.9.1 Amaranth*

The amaranthus belongs to Amaranthaceae family; it has more than 60 species which are grown across globe. *Amaranthus cardates* is one of those 60 species which is selected and used majorly for consumption. It is widely grown in Peru and other South American countries (Caballero et al., 2003). *Amaranthus cruentus* is grown in Guatenmala while *Amaranthus hypochondriacus* in Mexico. In India the pseudocereal of amaranth is widely consumed during fasting. It is very rich source of calcium, magnesium, and iron. It improves the hemoglobin in the blood. It is also high in protein content. The amino acids composition in the amaranthus is well balanced than other cereals and grains. The major parts of the protein present in amaranthus are albumin and globulin while prolamines are in lesser proportions. It also comprises with vitamins such as riboflavin, tocopherol (Ballabio et al., 2011; Chand

& Mihas, 2006). In the study conducted by (Gambus et al., 2002), the amaranthus flour was used in the preparation of bread to replace the gluten. At 10% replacement the nutritional properties were improved as compared to the standard composition while the sensory properties remained unaffected.

### 5.9.2 Quinoa

The quinoa botanically known as *Chenopodium quinoa* belongs to the amaranthus family. Quinoa is a pseudocereal which was a staple food from the ancient civilizations of Andes in South America. But now a ways it grown all across globe in all continents of Europe, Africa, and Asia etc. the quinoa has a similar appearance like oil seeds and possesses high oil content hence it can also be classified as pseudo-oil seed. The quinoa is available in different variety of colors such as white, red, purple, and black (Saturni et al., 2010; Vega-Gálvez et al., 2010; Bhargava et al., 2006). The proteins from the quinoa resemble very similar functional properties to that of milk protein. The quinoa protein completes the nutritional value since its protein possesses a high biological value of 83%. It is possible due the presence of combination of essential amino acids which provides it good functional properties. It also contains minerals in good proportions making it valuable for the human consumptions. The mainly found minerals are magnesium, manganese, iron, zinc and calcium (Repo-Carrasco et al., 2003; Vega-Gálvez et al., 2010). From the various studies conducted by researchers, food technologists (Zevallos et al., 2012; Mäkinen et al., 2013). It can be concluded that quinoa is safe for the replacement of gluten to produce and formulate gluten free product by using alternative proteins. The lower concentration of oat malt less than 1%, improved the bread volume and crumby structure (Mäkinen et al., 2013). The complete in vivo characterization of the developed gluten free product by using the alternative proteins should be done in order to understand its digestibility and the reactivity.

### 5.9.3 Buckwheat

The buck wheat belongs to polygonaceae family and caryophyllales order while botanically it is classified as fruit. But it is consumed in the form of grain or flour. The buck wheat is toasted before grinding it to flour. Since, the buck heat has reported some allergy cases in korea, Europe and japan (Panda et al., 2010). The toasting of buck wheat will denature the allergen compounds. There are two types/species of buckwheat which are widleychoosen for the human consumption namely *Fagopyrum esculentum* or common buckwheat and *Fagopyrum tartaricum*or tartary buckwheat. The *Fagopyrum tartaricum* is largely cultivated in Asian countries (Ikeda, 2002; Skrabanja et al., 2004). Buckwheat is a rich source for dietary fiber, vitamins, essential minerals, trace elements, rutin. The major factor for choosing the buckwheat for gluten replacement is its favorable amino acid sequence composition which is desirable to improve the protein quality of newly developed gluten free formulation for any product (Panda et al., 2010; Dunmire & Tierney, 1997). Hence buckwheat can be a suitable option to replace the gluten.

## 5.10 Functionality of Proteins

Apart from the gluten's textural properties and its effect, there are various other properties exhibited by the protein known as functional properties which govern the physical, chemical, organoleptic and nutritional properties of any product. Valuable dimensions have imparted by the functional properties of the protein to the various products in terms of its product's texture, appearance, taste as well as nutrition. The functionalities of protein include solubility, emulsification, foaming, water holding, and oil holding capacity, gelation, surface hydrophobicity, etc. which are described in detail as follows.

### *5.10.1 Solubility*

The index of protein functionality can be measured from the solubility which can be estimated by the aggregation and denaturation of protein. The improvement in the functionality of protein can be achieved by improving its solubility (Chobert et al., 1988a). Since some product require soluble protein and some require insoluble protein. The solubility of protein depends on the various factors such as temperature, pH, isolectric point etc (Chobert et al., 1988b). The solubility of protein plays an important role in the replacement of gluten in the products such as soup powders, instant soups and curries mixes which are having a major component of wheat flour and soy bean flour. In such products the protein solubility is desirable (Mutilangi et al., 1996). Hence such soluble protein can be replaced by the whey protein and whey protein isolates which are soluble in water. While the proteins in the bakery products must be insoluble to impart the textural and rheological properties. The gluten protein is responsible for the structure of bakery products (Garrett & Hunt, 1974). The caseinate protein from milk and some soy proteins which are insoluble can be used as an alternative for gluten in the bakery products . But before replacing the gluten by caseinate and soy protein its rheological properties are to be checked. Both the soluble and non-soluble proteins are important in respect to replace the gluten as per their utility in the final product possessing desirable properties.

### 5.10.2 Emulsification

Emulsification is a property where the oil water interfaces are hold together to form an emulsion. The emulsion quality of the protein is measured from its water-oil holding and binding (Turgeon et al., 1992). The interaction at the surface of oil and water at their surfaces are studied to know the emulsification property. The emulsifiers are measured for the emulsification activity, emulsion forming capacity and the stability of formed emulsion (Nakai et al., 1980). These properties of emulsion depend on the molecular structure of the protein molecule. The whey protein from milk and soybean protein, their molecular structure is different, hence both exhibits different emulsion properties. As a result they may find various applications as per the product suitability. The soy bean proteins are considered as one of the best proteins possessing best quality of emulsification quality, hence it can be used an alternative protein to replace the gluten.

### 5.10.3 Foaming

The foaming property is a unique property exhibited by protein generally in the products such as marshmallows and other edible foams. The molecular properties of the protein are related to the foaming characteristics (Kitabatake & Doi, 1982). The capacity to form foam and its stability are the important parameters to check these foam characteristics. The peptide linkages of amino acid sequences numbering from 101–145, 107–153, and 107–145 are important for the formation of superior quality of foam (German & Phillips, 1994). Foams are generally comprised of two phases a continuous aqueous phase and another dispersed gaseous phase. For obtaining food quality foam, segmental flexibility which is unfolding of bonds at the interfaces of the molecular secondary interactions of charged and polar groups is very important (Althouse et al., 1995). The soybean protein and whey protein are mainly used for their foaming property. The protein molecules present in these proteins possess a strong tendency to form the dimers, trimmers, and high order peptide- peptide linkages that are necessary for the generation of foam (Cheftel et al., 1985). Where gluten protein molecules have fewer tendencies to form such linkages and ultimately the low-quality foam is obtained. Hence, these alternative proteins can be used very well in place of the gluten for their better foaming characteristics.

### 5.10.4 Surface Hydophobicity

The functional properties of the protein are impacted by the structural characteristics which are used to evaluate the protein conformation while can be measured by the surface hydrophobicity of the protein molecules (Nakai et al., 1980). The protein molecules are susceptible to heat treatments, resulting into the surface hydophobicity. The heat treatments given to the protein molecules may result in the denaturation and hydrolysis giving rise to two separate effects (Kato et al., 1983) have studied these effects in the research conducted by him. In the research he concluded that, when the good co-relations are established between the surface tension, emulsifying activity, interfacial tension of the proteins present in the sunflower, rapeseed and soy bean. Due to the presence of the shorter amino acid chains and secondary structures, this surface hydophobicity is affected. The properties of the protein molecules are important for the formation of structures in the bakery products, foams, gels, confectionary products etc. the surface hydophobicity is an important parameter to study for the gluten and non-gluten proteins.

### 5.10.5 Gelation

The formation of gel by the entrapment of water molecule with the minimum synerisis is the gelation property which is generally referred to the structural strength of the product. When the protein-protein interaction increases with the water, it results in the formation of gel (Matsumoto & Hayashi, 1990). The products' mechanical strength and its viscoelastic properties can be determined from the gel strength. The gel formation and the gel strength may depend on the various extrinsic factors such as pH, ionic strength, temperature etc. The environmental conditions govern the gel characteristics; the characteristics includes such as firmness, rate of gelation, transparency of gel and its microstructure (Aguilera, 1995). The gel forming characteristics of the protein can be used as a replacement for the gluten. Many of the researchers have replaced the gluten by using various gums such as moringa seed gum, guar gum, gum arabica, gum gatti etc. Gum binds all the ingredients well (Kitabatake & Doi, 1993). But there are certain disadvantages of using the gum as replacement of gluten. The gum binds all ingredients but imparts stickiness to the product and increases the calorie content. Gum can't provide the desired springiness to the product. Hence protein can be suitable option to replace both gluten and gum. The protein possesses both the properties i.e. to hold and bind all ingredients together as well as it can impart the desired springiness to the product. The gel forming property of protein will improve the bonding and interactions; it will help to improve the texture of product. The protein will impart good structure and strength to the product. It also adds benefits by improving the protein content of the product.

### 5.10.6 *Water Holding Capacity*

The protein water interactions are one of the important characteristics which govern the water bounding and extent of water holding (Fevzioglu et al., 2012). The rheological properties of the bakery products are governed by the water holding capacity of protein molecules. The caseinate and β-lactoglobulin are water-insoluble proteins and possess some similar characteristics to that of gluten (Kinsella et al., 1989). The water holding capacity of protein influences the textural and structural properties of any product. When the water holding capacity of non-gluten protein increases and solubility decreases, such protein molecule can provide similar rheological properties as of gluten (Kneifel & Seiler, 1993). As discussed caseinate and β-lactoglobulin can be alternative to replace the gluten in the bakery products since desired functionality can be obtained.

### 5.10.7 *Oil Holding Capacity*

The oil holding property of protein is the most important property for the production of bakery products. Cakes, biscuits, crackers, cupcakes, donuts, muffins, etc. product requires fats as a compulsory ingredient to impart the softness. The fluffiness, springiness of these bakery products depends on the protein and the oil holding capacity of the protein. The protein molecules hold the oil in therewith to provide the desired softness. Oil holding plays an important role in the formation of emulsions, foams, and their stability (Gauthier et al., 1993). Hence when the gluten-free product has to be developed, it should be kept in mind the proteins from soybean can be alternative to gluten, since they possess more oil holding capacity. The oil holding capacity of proteins will govern the softness of the product as well it is also responsible for the flavor holding.

## 5.11 Strategies for Replacement of Gluten

When the gluten protein has to be replaced, several combinations of the protein from various sources can be used. The functional properties of the protein can be altered and modified to obtain the best results and superior quality products. The functionalities of these proteins can be altered by using different processes such as as enzymatic modification, high-pressure modification, cross linking of proteins, ultrasound treatment to proteins. These processes can be termed as strategies for the replacement of gluten and are discussed as follows.

### 5.11.1 Enzymatic Modification

The functionality of the proteins can be enhanced by the usage of enzymatic modification in order to achieve the desired properties, which can be used as alternative protein in place of gluten. The enzymatic modification is done by altering the amino acid chain sequence which will yield into the desired functionality of the specific protein. The enzymatic modification can be achieved by proteolysis, it is also termed as peptide linkage hydrolysis (Panyam & Kilara, 1996). The mechanism of the enzymatic modification can be seen by three different effects occurring in the protein molecule. There is decrease in the molecular weight of protein molecule, an increase in the ionizable groups while hydrophobic groups are concealed as a result of enzymatic modification (Kim et al., 1990). All of these results into the improved solubility reduced surface hydophobicity, increment in the gelation, foaming and emulsification capacity of the proteins. While carrying out enzymatic modification of the protein, the enzyme specificity is one of the key factors affecting the modification as well resulting in the faults in achieving the functionality of the protein (Whitaker, 1977). The other factors affecting the modification process are pH, isolectric point, ionic strength, protein denaturation, enzyme concentration, temperature. The molecular size of protein, specific amino acid sequence, structure of protein molecule will govern the functionality of protein (Casey et al., 1991). To achieve the desired functional properties, such as emulsification, foaming. Surface hydophobicity, water holding, oil holding and flavor binding capacity of the protein molecules can be modified enzymatically in order to utilize them as a replacement for gluten. The enzymatic modification of protein finds potential option to improve protein functionality which can be alternative to be used in place of gluten.

### 5.11.2 High Pressure Modification

The functionality of the protein can be modified by the application of high pressure. (Messens et al., 1997) have successfully carried research on the modification of the functional properties of proteins from the milk, egg, soy and meat proteins by the use of high pressure. The high pressure modification process carried at a pressure of 1000 mPa can confront the protein to modify resulting in desirable changes in its functional properties. The mechanism for modification can be explained as, since the high pressure is applied; the volume of the protein molecule decreases due to the compression of the internal cavities. It also results in the rupture of covalent bond interactions and establishment of new intra and inter molecular bonds within the protein molecules. As a result of these secondary, tertiary and quaternary protein molecules are stabilizes. In the research conducted by (Puppo et al., 2004) it was found that the hydophobicity of the protein can be reduced by using high pressure. The pressure of 150 mPa can be applied to stabilize the hydrophobic interactions between the protein molecules. This will contribute to enhancing the water holding capacity of the protein. 200 mPa pressure can be used for the hydrophobic interaction stabilization of secondary, tertiary protein molecules. From the study of (Singh & Ramaswamy, 2015) it was included that the β-lactoglobulin present in the casein-a milk protein can be modified for its gel-forming capacity. The interactions between the β-lactoglobulin and α-lactoglobulin can be improved; resulting in an improved gel-forming capacity. The high pressure can also be applied to whey proteins as well. The emulsification efficiency of soy protein has been improved due to the application of high pressure.

### 5.11.3 Cross-Linking of Proteins

The food biopolymers play an important role in the formation of effective structure in the bakery products. The cross-linking of protein is an important tool to alter the structure of the protein molecule by which the functional; properties will be enhanced. The cross-linkage of protein molecules may be achieved by the enzymatic treatment and non-enzymatic Maillard reaction. The enzymes used for the cross-linking are transglutaminase, tyrosinase, lactases, peroxidase, and sulfhydryl oxidase (Buchert et al., 2010). The specificity of cross-linking the proteins is affected by the enzymatic catalyst and the mild reaction conditions. The reaction condition includes temperature, pH, isolectric points etc. the disulphide linkages govern the thermo-rheological properties of dough (Gerrard, 2006). The cross linking can be achieved by the heat treatment, enzyme treatment, ultrasound treatment, which will create the intentional inter and intra molecular bonding of sulfhydryl groups of cysteine in the protein molecules. The creation of cross linking may lead to enhanced stability of the protein and will resist to proteolysis (Matheis & Whitaker, 1987). The cross linking of the disulphide's in the protein will strengthen the stability, firmness of protein and also enhances the viscoelastic properties. The cross linking of the proteins may also enhance the digestibility and reduces the allergenicity of the protein (Thalmann & Lötzbeyer, 2002). The non–globular proteins may be more susceptible for the cross linking while the cross linking in globular proteins may enhance its properties and functionality may be altered. The modified protein by cross linking may be used in place of gluten for development of gluten free product. The modified protein by cross linking will exhibit the similar characteristics to that of gluten due the cross linking and network formation. The cross-linking is a cost-effective and rapid tool which can be used for the modification of protein with good efficacy.

### 5.11.4 Ultrasound Treatment to Proteins

Ultrasound treatment is a new, safe, and one of the effective methods utilized for the modification of the functionality of the protein. The mechanism of the ultrasound modification is based on the shear stresses and turbulence created by the high-intensity ultrasound (Jambrak et al., 2009). The ultrasound treatment given to soybean protein has to yield good results for the viscosity and elastic characteristics which were recorded on the stress rheometer by (Arzeni et al., 2012). The impact of high-intensity ultrasound treatment was studied by (Riener et al., 2009). In the study conducted by (Mishra et al., 2001) egg protein, whey protein, and whey protein isolate were treated at 750 W, 20 kHz frequency and 20% amplitude. The satisfactory changes in the gelation, solubility and viscosity are closely related to the modification done on the molecular level of the protein molecule. Thus the functional properties of the protein are widely affected by the ultrasound treatment which makes ultrasound as a potential option to alter the functionality of protein and making it suitable for the further development of the desired gluten free product.

## 5.12 Gluten Free Products

The gluten free product available in market are enlisted in the Table 5.1, comprising with the alternative proteins.

## 5.13 Challenges for Gluten Replacement

When gluten is replaced from any product, it is necessary to check the functional, nutritional and organoleptic properties of the product and protein. The gluten protein influences all of these properties of the product. Hence the replacement of gluten by an alternative protein can face these challenges, which are discussed as follows.

### 5.13.1 Nutritional Challenges

Now a day's consumers are aware of the product they are consuming. Whether the product is nutritionally rich, does it contains essential nutrients is checked by the consumer before purchasing it. Since the gluten is replaced from the products, it has created nutritional imbalance. Limited numbers of evidences are available regarding its nutritional inadequacies (Kinsella, 1982). These nutritional deficiencies can be fulfilled by the combination of two or more proteins from varying sources. These

**Table 5.1** Gluten free products prepared from alternative proteins

| S.N. | Product name | Alternative protein | References |
|---|---|---|---|
| 1. | Bread | Whey protein | Storck et al. (2013) |
| | | Soybean protein isolates | Crockett et al. (2011) |
| | | Maize protein | Fevzioglu et al. (2012) |
| | | Albumin | Schoenlechner et al. (2010) |
| | | Soy protein isolates, Pea protein, lupine | Ziobro et al. (2013) |
| | | Carob flour (Caroubin) | Tsatsaragkou et al. (2014) |
| 2. | Pasta | Casein + Egg whites | Sozer (2009) |
| | | Egg albumin | Marti et al. (2014) |
| | | Whey protein | Kumar et al. (2019) |
| | | Whey protein+ Grape peel+ corn starch | Ungureanu-Iuga et al. (2020) |
| 3. | Cupcake | Amaranth + egg protein | Egorova and Reznichenko (2018) |
| 4. | White sauce | Soy bean protein | Gularte et al. (2012) |
| 5. | Muffins | Pea protein, Soy protein isolates | Matos et al. (2014) |
| | | Casein | Hubbell et al. (2007) |
| | | Chick pea protein | Herranz et al. (2016) |
| 6. | Donuts | Egg protein + Whey + casein | Melito and Farkas (2013) |
| 7. | Bakery products | Peas protein isolates | Mariotti et al. (2009) |
| 8. | Fermented products | Pea protein + soy bean protein isolates | Marco and Rosell (2008) |
| 9. | Crackers | Buckwheat + protein isolates | Sedej et al. (2011) |
| 10. | Noodles | Zein + Rice flour | Jeong et al. (2017) |

combinations of the alternative proteins which have replaced the gluten are having some of the important essential amino acids such as lysine, and tryptophan. (Thompson, 1999), which are missing in the gluten fractions. Thus the new alternative protein combinations can fulfill the nutritional demands of the protein. Along with it, gluten-free products can be fortified with iron, niacin, folate, riboflavin, and thiamine (Thompson, 2000). The replacement of gluten by the combination of proteins brings a challenge to keep the glycemic index or the calorie count in the limit. If the glycemic index or the calorie goes beyond the limits, the replacement is of no use.

### *5.13.2 Functional Challenges*

The gluten protein provides the essential structure building property to the product. This structural property of gluten is one of its important functional properties. It also governs the visco-elasticity of the dough. The gliadin and gluten in ratio present in the gluten are responsible for these functional properties. In gluten-free products, it is a major challenge to replace gluten by the alternative protein to provide similar resembling functional properties (Smith et al., 2014). The loaf volume of bread, softness of cakes, muffins and cupcakes, the springiness of noodles, spaghetti, and pasta; all of these are the functional properties imparted by gluten. The consumer is habitual to consume these products due to its functional properties. If the gluten is replaced by the alternative protein and it does not meet the similar characteristics of functional properties imparted by gluten, the product will fail and will not meet consumer satisfaction.

### *5.13.3 Organoleptic Challenges*

When the gluten-free product is developed, organoleptic factors should be always considered to obtain a product which is having similar properties to that of gluten-containing products. The most important property imparted by the gluten is its texture; the people are habitual to the texture of gluten imparted in various products such as cakes, biscuits, bread, cookies, etc. (De Wit, 1998). The organoleptic characteristics of any product are judged by color, flavor, texture, appearance, taste, etc. Making the gluten free product of the similar sensory characteristics is one of the most important organoleptic challenges faced by the researcher (De Wit, 1990). Where the consumer has an impact of the taste, texture, color and flavor of the gluten based product, denies the acceptance of the product. The perfect combination of the alternative source of protein which will provide perfect texture, structure, product appearance, taste, color is most important and difficult challenge faced by the researchers. In case of breads, the gluten free bread should possess same loaf volume as that of gluten containing; the gluten free cakes should impart same softness as that of gluten containing cakes; the gluten free sauces, and premixes should provide same consistency, visco-elastic properties and appearance to that of gluten containing.

## 5.14 Conclusion

The development of gluten-free products by using alternative protein is not a straight and easy process. It involves many steps and complex processes. Many of the food technologist, researchers, and scientists are working on the molecular chemistry of the non-gluten proteins to produce a perfect combination which will impart the functional as well as nutritional properties to the newly developed gluten-free product. In products such as breads, biscuits, cookies, and cakes many of the researchers have found out a perfect combination of alternative proteins which are providing similar characteristics to gluten in terms of texture, structure of product, taste, overall appearance and enhanced with more nutritional properties.

# References


Aghamirzaei, M., Heydari-Dalfard, A., Karami, F., & Fathi, M. (2013). Pseudo-cereals as a functional ingredient: Effects on bread nutritional and physiological properties-review. *International Journal of Agriculture and Crop Sciences (IJACS), 5*(14), 1574–1580.

Aguilar, N., Albanell, E., Miñarro, B., & Capellas, M. (2015). Chickpea and tiger nut flours as alternatives to emulsifier and shortening in gluten-free bread. *LWT- Food Science and Technology, 62*(1), 225–232.

Aguilera, J. M. (1995). Gelation of whey proteins: Chemical and rheological changes during phase transition in food. *Food Technology (Chicago), 49*(10), 83–89.

Althouse, P., Dinakar, P., & Kilara, A. (1995). Screening of proteolytic enzymes to enhance foaming of whey protein isolates. *Journal of Food Science, 60*(5), 1110–1112.

Alvarez-Jubete, L., Auty, M., Arendt, E. K., & Gallagher, E. (2010). Baking properties and microstructure of pseudocereal flours in gluten-free bread formulations. *European Food Research and Technology, 230*(3), 437.

Andersson, H., Öhgren, C., Johansson, D., Kniola, M., & Stading, M. (2011). Extensional flow, viscoelasticity and baking performance of gluten-free zein-starch doughs supplemented with hydrocolloids. *Food Hydrocolloids, 25*(6), 1587–1595.

Arzeni, C., Martínez, K., Zema, P., Arias, A., Pérez, O., & Pilosof, A. (2012). Comparative study of high intensity ultrasound effects on food proteins functionality. *Journal of Food Engineering, 108*(3), 463–472.

Ballabio, C., Uberti, F., Di Lorenzo, C., Brandolini, A., Penas, E., & Restani, P. (2011). Biochemical and immunochemical characterization of different varieties of amaranth (Amaranthus L. ssp.) as a safe ingredient for gluten-free products. *Journal of Agricultural and Food Chemistry, 59*(24), 12969–12974.

Bhargava, A., Shukla, S., & Ohri, D. (2006). Chenopodium quinoa – An Indian perspective. *Industrial Crops and Products, 23*(1), 73–87.

Boye, J., Zare, F., & Pletch, A. (2010). Pulse proteins: Processing, characterization, functional properties and applications in food and feed. *Food Research International, 43*(2), 414–431.

Buchert, J., Ercili Cura, D., Ma, H., Gasparetti, C., Monogioudi, E., Faccio, G., Mattinen, M., Boer, H., Partanen, R., & Selinheimo, E. (2010). Crosslinking food proteins for improved functionality. *Annual Review of Food Science and Technology, 1*, 113–138.

Caballero, B., Trugo, L. C., & Finglas, P. M. (2003). *Encyclopedia of food sciences and nutrition*. Academic.

Casey, P. J., Thissen, J. A., & Moomaw, J. F. (1991). Enzymatic modification of proteins with a geranylgeranyl isoprenoid. *Proceedings of the National Academy of Sciences, 88*(19), 8631–8635.

Chan, P. S.-K., Chen, J., Ettelaie, R., Law, Z., Alevisopoulos, S., Day, E., & Smith, S. (2007). Study of the shear and extensional rheology of casein, waxy maize starch and their mixtures. *Food Hydrocolloids, 21*(5–6), 716–725.

Chand, N., & Mihas, A. A. (2006). Celiac disease: Current concepts in diagnosis and treatment. *Journal of Clinical Gastroenterology, 40*(1), 3–14.

Cheftel, J., Cuq, J., & Lorient, D., 2nd. (1985). In O. R. Fennema (Ed.), *Food chemistry*. Marcel Dekker.

Chobert, J. M., Bertrand-Harb, C., & Nicolas, M. G. (1988a). Solubility and emulsifying properties of caseins and whey proteins modified enzymically by trypsin. *Journal of Agricultural and Food Chemistry, 36*(5), 883–892.

Chobert, J. M., Sitohy, M. Z., & Whitaker, J. R. (1988b). Solubility and emulsifying properties of caseins modified enzymatically by Staphylococcus aureus V8 protease. *Journal of Agricultural and Food Chemistry, 36*(1), 220–224.

Comino, I., de Lourdes Moreno, M., Real, A., Rodríguez-Herrera, A., Barro, F., & Sousa, C. (2013). The gluten-free diet: Testing alternative cereals tolerated by celiac patients. *Nutrients, 5*(10), 4250–4268.

Crockett, R., Ie, P., & Vodovotz, Y. (2011). Effects of soy protein isolate and egg white solids on the physicochemical properties of gluten-free bread. *Food Chemistry, 129*(1), 84–91.

De Mesa-Stonestreet, N. J., Alavi, S., & Bean, S. R. (2010). Sorghum proteins: The concentration, isolation, modification, and food applications of kafirins. *Journal of Food Science, 75*(5), R90–R104.

De Wit, J. (1990). Thermal stability and functionality of whey proteins. *Journal of Dairy Science, 73*(12), 3602–3612.

De Wit, J. (1998). Nutritional and functional characteristics of whey proteins in food products. *Journal of Dairy Science, 81*(3), 597–608.

Deora, N. S., Deswal, A., Dwivedi, M., & Mishra, H. N. (2014). Prevalence of coeliac disease in India: A mini review. *International Journal of Latest Research Science Technology, 3*(10), 58–60.

Deora, N. S., Deswal, A., & Mishra, H. N. (2015). Functionality of alternative protein in gluten-free product development. *Food Science and Technology International, 21*(5), 364–379. https://doi.org/10.1177/1082013214538984

Dunmire, W. W., & Tierney, G. D. (1997). *Wild plants and native peoples of the Four Corners*. Museum of New Mexico Press.

Egorova, E., & Reznichenko, I. (2018). Development of food concentrate–semi-finished product with amaranth flour for gluten-free cupcakes. *Tekhnika i Tekhnologiya Pishchevykh Proizvodstv, 48*(2).

Esen, A. (1987). A proposed nomenclature for the alcohol-soluble proteins (zeins) of maize (Zea mays L.). *Journal of Cereal Science, 5*(2), 117–128.

Feillet, P., & Roulland, T. M. (1998). Caroubin: A gluten-like protein isolated from carob bean germ. *Cereal Chemistry, 75*(4), 488–492.

Fevzioglu, M., Hamaker, B. R., & Campanella, O. H. (2012). Gliadin and zein show similar and improved rheological behavior when mixed with high molecular weight glutenin. *Journal of Cereal Science, 55*(3), 265–271.

Foschia, M., Horstmann, S. W., Arendt, E. K., & Zannini, E. (2017). Legumes as functional ingredients in gluten-free bakery and pasta products. *Annual Review of Food Science and Technology, 8*, 75–96.

Gambus, H., Gambus, F., & Sabat, R. (2002). The research on quality improvement of gluten-free bread by amaranthus flour addition. *Zywnosc, 9*(2), 99–112.

Garrett, E. R., & Hunt, C. A. (1974). Physicochemical properties, solubility, and protein binding of Δ9-tetrahydrocannabinol. *Journal of Pharmaceutical Sciences, 63*(7), 1056–1064.

Gauthier, S., Paquin, P., Pouliot, Y., & Turgeon, S. (1993). Surface activity and related functional properties of peptides obtained from whey proteins. *Journal of Dairy Science, 76*(1), 321–328.

German, J. B., & Phillips, L. (1994). Protein interactions in foams: Protein-gas phase interactions. *Protein Functionality in Food Systems, 9*, 181–208.

Gerrard, J. (2006). Protein cross-linking in food. *Food Biochemistry and Food Processing*, 223.

Gianazza, E., Viglienghi, V., Righetti, P. G., Salamini, F., & Soave, C. (1977). Amino acid composition of zein molecular components. *Phytochemistry, 16*(3), 315–317.

Gujral, N., Freeman, H. J., & Thomson, A. B. (2012). Celiac disease: Prevalence, diagnosis, pathogenesis and treatment. *World journal of gastroenterology: WJG, 18*(42), 6036.

Gularte, M. A., Gómez, M., & Rosell, C. M. (2012). Impact of legume flours on quality and in vitro digestibility of starch and protein from gluten-free cakes. *Food and Bioprocess Technology, 5*(8), 3142–3150.

Herranz, B., Canet, W., Jiménez, M. J., Fuentes, R., & Alvarez, M. D. (2016). Characterisation of chickpea flour-based gluten-free batters and muffins with added biopolymers: Rheological, physical and sensory properties. *International Journal of Food Science & Technology, 51*(5), 1087–1098.

Hubbell, L., Goodwin, D., Samuel, L., & Navder, K. (2007). Effect of soy protein isolate on the textural and physical attributes of gluten-free muffins made from sorghum flour. *Journal of the American Dietetic Association, 107*(8), A76.

Ikeda, K. (2002). Buckwheat composition, chemistry, and processing. *Advance Food Nutritional Research, 44*, 395.
Jambrak, A. R., Lelas, V., Mason, T. J., Krešić, G., & Badanjak, M. (2009). Physical properties of ultrasound treated soy proteins. *Journal of Food Engineering, 93*(4), 386–393.
Jeong, S., Kim, M., Yoon, M.-R., & Lee, S. (2017). Preparation and characterization of gluten-free sheeted doughs and noodles with zein and rice flour containing different amylose contents. *Journal of Cereal Science, 75*, 138–142.
Jerome, R. E., Singh, S. K., & Dwivedi, M. (2019). Process analytical technology for bakery industry: A review. *Journal of Food Process Engineering, 42*(5), e13143.
Kato, A., Osako, Y., Matsudomi, N., & Kobayashi, K. (1983). Changes in the emulsifying and foaming properties of proteins during heat denaturation. *Agricultural and Biological Chemistry, 47*(1), 33–37.
Kaur, M., & Singh, N. (2005). Studies on functional, thermal and pasting properties of flours from different chickpea (Cicer arietinum L.) cultivars. *Food Chemistry, 91*(3), 403–411.
Kaur, M., & Singh, N. (2007). Characterization of protein isolates from different Indian chickpea (Cicer arietinum L.) cultivars. *Food Chemistry, 102*(1), 366–374.
Kim, S. Y., Park, P. S., & Rhee, K. C. (1990). Functional properties of proteolytic enzyme modified soy protein isolate. *Journal of Agricultural and Food Chemistry, 38*(3), 651–656.
Kinsella, J. (1982). Relationships between structure and functional properties of food proteins. *Food Proteins, 1*, 51–103.
Kinsella, J., Whitehead, D., Brady, J., & Bringe, N. (1989). Milk proteins: Possible relationships of structure and function. *Developments in Dairy Chemistry, 4*, 55–95.
Kitabatake, N., & Doi, E. (1982). Surface tension and foaming of protein solutions. *Journal of Food Science, 47*(4), 1218–1221.
Kitabatake, N., & Doi, E. (1993). Improvement of protein gel by physical and enzymatic treatment. *Food Reviews International, 9*(4), 445–471.
Kneifel, W., & Seiler, A. (1993). Water-holding properties of milk protein products-A review. *Food Structure, 12*(3), 3.
Kumar, C. M., Sabikhi, L., Singh, A., Raju, P., Kumar, R., & Sharma, R. (2019). Effect of incorporation of sodium caseinate, whey protein concentrate and transglutaminase on the properties of depigmented pearl millet based gluten free pasta. *LWT, 103*, 19–26.
Lending, C. R., & Larkins, B. A. (1989). Changes in the zein composition of protein bodies during maize endosperm development. *The Plant Cell, 1*(10), 1011–1023.
Liang, L., & Luo, Y. (2020). Casein and pectin: Structures, interactions, and applications. *Trends in Food Science & Technology, 97*, 391–403.
Luo, Y., Pan, K., & Zhong, Q. (2014). Physical, chemical and biochemical properties of casein hydrolyzed by three proteases: Partial characterizations. *Food Chemistry, 155*, 146–154.
Lupano, C. (2003). Discs for "empanadas" with whey protein concentrate. *Journal of Food Technology, 1*(4), 182–186.
Mäkinen, O. E., Zannini, E., & Arendt, E. K. (2013). Germination of oat and quinoa and evaluation of the malts as gluten free baking ingredients. *Plant Foods for Human Nutrition, 68*(1), 90–95.
Marco, C., & Rosell, C. M. (2008). Functional and rheological properties of protein enriched gluten free composite flours. *Journal of Food Engineering, 88*(1), 94–103.
Marcoa, C., & Rosell, C. M. (2008). Effect of different protein isolates and transglutaminase on rice flour properties. *Journal of Food Engineering, 84*(1), 132–139.
Mariotti, M., Lucisano, M., Pagani, M. A., & Ng, P. K. (2009). The role of corn starch, amaranth flour, pea isolate, and Psyllium flour on the rheological properties and the ultrastructure of gluten-free doughs. *Food Research International, 42*(8), 963–975.
Marston, K., Khouryieh, H., & Aramouni, F. (2016). Effect of heat treatment of sorghum flour on the functional properties of gluten-free bread and cake. *LWT- Food Science and Technology, 65*, 637–644.

Marti, A., Barbiroli, A., Marengo, M., Fongaro, L., Iametti, S., & Pagani, M. A. (2014). Structuring and texturing gluten-free pasta: Egg albumen or whey proteins? *European Food Research and Technology, 238*(2), 217–224.

Matheis, G., & Whitaker, J. R. (1987). A review: Enzymatic cross-linking of proteins applicable to foods. *Journal of Food Biochemistry, 11*(4), 309–327.

Matos, M. E., Sanz, T., & Rosell, C. M. (2014). Establishing the function of proteins on the rheological and quality properties of rice based gluten free muffins. *Food Hydrocolloids, 35*, 150–158.

Matsumoto, T., & Hayashi, R. (1990). Properties of pressure-induced gels of various soy protein products. *Nippon Nogeikagaku Kaishi, 64*(9), 1455–1459.

Melito, H., & Farkas, B. E. (2013). Physical properties of gluten-free donuts. *Journal of Food Quality, 36*(1), 32–40.

Merga, B., & Haji, J. (2019). Economic importance of chickpea: Production, value, and world trade. *Cogent Food & Agriculture, 5*(1), 1615718.

Messens, W., Van Camp, J., & Huyghebaert, A. (1997). The use of high pressure to modify the functionality of food proteins. *Trends in Food Science & Technology, 8*(4), 107–112.

Miñarro, B., Albanell, E., Aguilar, N., Guamis, B., & Capellas, M. (2012). Effect of legume flours on baking characteristics of gluten-free bread. *Journal of Cereal Science, 56*(2), 476–481.

Mishra, S., Mann, B., & Joshi, V. (2001). Functional improvement of whey protein concentrate on interaction with pectin. *Food Hydrocolloids, 15*(1), 9–15.

Mutilangi, W., Panyam, D., & Kilara, A. (1996). Functional properties of hydrolysates from proteolysis of heat-denatured whey protein isolate. *Journal of Food Science, 61*(2), 270–275.

Nakai, S., Ho, L., Helbig, N., Kato, A., & Tung, M. (1980). Relationship between hydrophobicity and emulsifying properties of some plant proteins. *Canadian Institute of Food Science and Technology Journal, 13*(1), 23–27.

Oom, A., Pettersson, A., Taylor, J. R., & Stading, M. (2008). Rheological properties of kafirin and zein prolamins. *Journal of Cereal Science, 47*(1), 109–116.

Panda, R., Taylor, S. L., & Goodman, R. E. (2010). Development of a sandwich enzyme-linked immunosorbent assay (ELISA) for detection of buckwheat residues in food. *Journal of Food Science, 75*(6), T110–T117.

Panyam, D., & Kilara, A. (1996). Enhancing the functionality of food proteins by enzymatic modification. *Trends in Food Science & Technology, 7*(4), 120–125.

Pico, J., Reguilón, M. P., Bernal, J., & Gómez, M. (2019). Effect of rice, pea, egg white and whey proteins on crust quality of rice flour-corn starch based gluten-free breads. *Journal of Cereal Science, 86*, 92–101.

Pradhan, D., Hoque, M., Singh, S. K., & Dwivedi, M. (2021). Application of D-optimal mixture design and artificial neural network in optimizing the composition of flours for preparation of Gluten-Free bread: Optimization of ingredient for preparation of gluten free bread. *Journal of microbiology, biotechnology and food sciences*, e3294–e3294.

Puppo, C., Chapleau, N., Speroni, F., de Lamballerie-Anton, M., Michel, F., Añón, C., & Anton, M. (2004). Physicochemical modifications of high-pressure-treated soybean protein isolates. *Journal of Agricultural and Food Chemistry, 52*(6), 1564–1571.

Ranhotra, G., Gelroth, J., Glaser, B., Lorenz, K., & Johnson, D. (1993). Composition and protein nutritional quality of quinoa. *Cereal Chemistry, 70*, 303–303.

Repo-Carrasco, R., Espinoza, C., & Jacobsen, S.-E. (2003). Nutritional value and use of the Andean crops quinoa (Chenopodium quinoa) and kañiwa (Chenopodium pallidicaule). *Food Reviews International, 19*(1-2), 179–189.

Resch, J., & Daubert, C. (2002). Rheological and physicochemical properties of derivatized whey protein concentrate powders. *International Journal of Food Properties, 5*(2), 419–434.

Riener, J., Noci, F., Cronin, D. A., Morgan, D. J., & Lyng, J. G. (2009). Characterisation of volatile compounds generated in milk by high intensity ultrasound. *International Dairy Journal, 19*(4), 269–272.

Saturni, L., Ferretti, G., & Bacchetti, T. (2010). The gluten-free diet: Safety and nutritional quality. *Nutrients, 2*(1), 16–34.

Schoenlechner, R., Drausinger, J., Ottenschlaeger, V., Jurackova, K., & Berghofer, E. (2010). Functional properties of gluten-free pasta produced from amaranth, quinoa and buckwheat. *Plant Foods for Human Nutrition, 65*(4), 339–349.

Sedej, I., Sakač, M., Mandić, A., Mišan, A., Pestorić, M., Šimurina, O., & Čanadanović-Brunet, J. (2011). Quality assessment of gluten-free crackers based on buckwheat flour. *LWT- Food Science and Technology, 44*(3), 694–699.

Sharma, D., Gupta, R., & Joshi, I. (2014). Nutrient analysis of raw and processed soybean and development of value added soybean noodle. *Inventory Journal, 1*, 1–5.

Singh, A., & Ramaswamy, H. (2015). High pressure modification of egg components: Exploration of calorimetric, structural and functional characteristics. *Innovative Food Science & Emerging Technologies, 32*, 45–55.

Skrabanja, V., Kreft, I., Golob, T., Modic, M., Ikeda, S., Ikeda, K., Kreft, S., Bonafaccia, G., Knapp, M., & Kosmelj, K. (2004). Nutrient content in buckwheat milling fractions. *Cereal Chemistry, 81*(2), 172–176.

Smith, B., Bean, S., Herald, T., & Aramouni, F. (2012). Effect of HPMC on the quality of wheat-free bread made from carob germ flour-starch mixtures. *Journal of Food Science, 77*(6), C684–C689.

Smith, B. M., Bean, S. R., Selling, G., Sessa, D., & Aramouni, F. M. (2014). Role of non-covalent interactions in the production of visco-elastic material from zein. *Food Chemistry, 147*, 230–238.

Sozer, N. (2009). Rheological properties of rice pasta dough supplemented with proteins and gums. *Food Hydrocolloids, 23*(3), 849–855.

Srikanlaya, C., Therdthai, N., Ritthiruangdej, P., & Zhou, W. (2018). Effect of hydroxypropyl methylcellulose, whey protein concentrate and soy protein isolate enrichment on characteristics of gluten-free rice dough and bread. *International Journal of Food Science & Technology, 53*(7), 1760–1770.

Storck, C. R., da Rosa Zavareze, E., Gularte, M. A., Elias, M. C., Rosell, C. M., & Dias, A. R. G. (2013). Protein enrichment and its effects on gluten-free bread characteristics. *LWT-Food Science and Technology, 53*(1), 346–354.

Thalmann, C., & Lötzbeyer, T. (2002). Enzymatic cross-linking of proteins with tyrosinase. *European Food Research and Technology, 214*(4), 276–281.

Thompson, T. (1999). Thiamin, riboflavin, and niacin contents of the gluten-free diet: Is there cause for concern? *Journal of the Academy of Nutrition and Dietetics, 99*(7), 858.

Thompson, T. (2000). Folate, iron, and dietary fiber contents of the gluten-free diet. *Journal of the Academy of Nutrition and Dietetics, 100*(11), 1389.

Tsatsaragkou, K., Yiannopoulos, S., Kontogiorgi, A., Poulli, E., Krokida, M., & Mandala, I. (2014). Effect of carob flour addition on the rheological properties of gluten-free breads. *Food and Bioprocess Technology, 7*(3), 868–876.

Turgeon, S., Gauthier, S., & Paquin, P. (1992). Emulsifying property of whey peptide fractions as a function of pH and ionic strength. *Journal of Food Science, 57*(3), 601–604.

Ungureanu-Iuga, M., Dimian, M., & Mironeasa, S. (2020). Development and quality evaluation of gluten-free pasta with grape peels and whey powders. *LWT, 130*, 109714.

van Riemsdijk, L. E., van der Goot, A. J., Hamer, R. J., & Boom, R. M. (2011). Preparation of gluten-free bread using a meso-structured whey protein particle system. *Journal of Cereal Science, 53*(3), 355–361.

Vega-Gálvez, A., Miranda, M., Vergara, J., Uribe, E., Puente, L., & Martínez, E. A. (2010). Nutrition facts and functional potential of quinoa (Chenopodium quinoa willd.), an ancient Andean grain: A review. *Journal of the Science of Food and Agriculture, 90*(15), 2541–2547.

Wieser, H. (1996). Relation between gliadin structure and coeliac toxicity. *Acta Paediatrica, 85*, 3–9.

Wthakar. (1977). *Enzymatic modification of proteins applicable to foods*. ACS Publications.

Zevallos, V. F., Ellis, H. J., Šuligoj, T., Herencia, L. I., & Ciclitira, P. J. (2012). Variable activation of immune response by quinoa (Chenopodium quinoa Willd.) prolamins in celiac disease. *The American Journal of Clinical Nutrition, 96*(2), 337–344.

Ziobro, R., Witczak, T., Juszczak, L., & Korus, J. (2013). Supplementation of gluten-free bread with non-gluten proteins. Effect on dough rheological properties and bread characteristic. *Food Hydrocolloids, 32*(2), 213–220.

# Chapter 6
# Regulatory and Labelling

**Murakonda Sahithi and Madhuresh Dwivedi**

**Abstract** Gluten in food is the major health concern for Gluten-intolerant people worldwide for many years. This can be minimized by the avoidance of gluten in the diet as it does not have a cure. The control and management of gluten in food have been recognized by many countries as an important food safety concern and risk management problem in recent years. Many nation governments and organizations have implemented laws, policies, regulations for the manufacturers, retailers, and marketers to indicate the label "gluten-free" on the package to communicate, control and manage the existence or non-existence of gluten in food. The regulations for gluten-free are designed to specify no adverse level of gluten (below 20 ppm) than fully gluten absence. Numerous countries such as South Africa, the united states, Canada, Europe, China, Japan, India, and Australia, etc., have issued "Gluten-free" regulations. The regulation for "Gluten-free foods" differs from the country. The public should know the regulations and labels of the organization to avoid the mis-consumption of "Gluten-free "foodstuffs". The word "Gluten-free" on labeling is provided by quantitatively detecting the gluten content by standardizing methods of Analysis like enzyme-linked immunoassay (ELISA). The regulations are the corporative linkage between the government, industry, and consumer.

**Keywords** Gluten-free · ELISA · Retailer

## 6.1 Introduction

The significance of diet and nutrition in food to prevent numerous health concerns was not established up to many centuries. The toxicity caused by food allergen-gluten and gluten intolerance in people leads to health issues such as celiac disease, dermatitis, gluten ataxia, herpetiformis, intestinal problems which leads to anemia, diabetes, cancer has been recognized internationally in recent years and became

M. Sahithi · M. Dwivedi (✉)
Department of Food Process Engineering, NIT Rourkela, Rourkela, Odisha, India
e-mail: dwivedim@nitrkl.ac.in



N. Singh Deora et al. (eds.), *Challenges and Potential Solutions in Gluten Free Product Development*, Food Engineering Series,
https://doi.org/10.1007/978-3-030-88697-4_6

serious health concern worldwide (Ahmad et al., 2019; Cornicelli et al., 2018; Tovoli et al., 2015). Moreover, these health issues directly cause an economic burden and indirectly causes loss of cost due to less productivity and more diseases. The major health concern is gluten in food for many people which can be minimized by avoidance of gluten in the diet as it does not have a cure (Sharma et al., 2015). Even though gluten-free food is manufactured, the contamination i.e. food adulteration can be caused anywhere from farm to fork-like milling, processing, cross-contact machinery during manufacturing, retail outlets, common utensils used in households ultimately leads to a health concern (Dudeja et al., 2016; Tripathi et al., 2017; Rifna et al., 2019).

The control and management of gluten in food have become an important food safety concern in recent years. Hence, it's essential to enable the consumer to differentiate between gluten and gluten-free food and make health preferences. The clear indication of the presence of gluten and gluten-free in certain foods is required for gluten-intolerant people. This felt the need to start the regulatory process to achieve food safety by determining of an optimum standard of ingredient content in the food. The awareness of the need of regulation and food allergens came in period 1990s, but the implementation of regulation globally came in the 2000s (Astley, 2019; Lee et al., 2014). The regulations assist the manufacturer to communicate, control and manage the existence or non-existence of gluten in food. The numerous countries such as South Africa, united states, Canada, Europe, China, Japan, India, UK and Australia etc. has recognized the importance of the polices, regulations and standard for the safety of gluten-sensitive people (Haraszi et al., 2011).The regulation of food label differs with the country. The regulations are established to certify that the anyallegation on a food label is obvious and verified by scientific verification (Duttaroy, 2019). The regulation for labeling requirement has also been mandatory for the awareness of ingredients causing allergies as it provides the consumer the information to communicate and choose whether it is suitable for consumption or not (Popping & Diaz-Amigo, 2018). The regulation aims to provide the assurance of consumer protection about food information. The consumer has the opportunity of opting between the bundle of quality attributes labeled on the package (Martini et al., 2019; Rostami et al., 2017; Sainsbury et al., 2018).

Furthermore, different organizations have been developed for identifying the approaches for gluten-free standards and set label regulations. These organizations have focused on delivering information to avoid food that triggers gluten reaction in gluten-tolerant people (Gendel, 2012; Madhuresh et al., 2013). The regulations are the corporative linkage between the government, industry, and consumer (Desmarchelier & Szabo, 2008). The chapter showcases the regulations standards of different organizations and labeling requirements for gluten-free food.

## 6.2 Regulatory and Labelling for "Gluten-Free" Foods

Regulations for "gluten-free" are designed to indicate no adverse level of gluten other than fully gluten absence (Akobeng, 2008). The regulation enforcing regarding the term Gluten-free on labeling is provided by quantitatively detecting the gluten content by standardizing methods of Analysis. The reliable methods for quantification and detection of gluten are mandatory to certify the safety of gluten-free foods. There are many methods for quantification and detection of gluten like enzyme-linked immunoassay (ELISA) which is mostly accepted technique as per regulations followed by many countries and other detection methods include polymer chain reaction (PCR) for DNA quantification, MS for gluten protein detection, HPLC, potentiometric electron tongue, optical biosensor, etc., (Alimentarius, 2003; Laube et al., 2011; Rosell et al., 2014; Sharma et al., 2015; Thompson & Méndez, 2008).

The standards and regulations set by national and international bodies should be strictly followed while developing or manufacturing gluten-free food. The regulations have been adopted for many products such as snacks, baked products, Extruded products, food especially processed to reduce gluten content, restaurant food, etc., for gluten-free foods (Ahmad et al., 2019; Navarro et al., 2017; Jerome et al., 2019). The regulatory standards and policies set for the Gluten-free products need to be followed along with the existing well-recognized food safety management bodies like "Good manufacture practices (GMP)" and "Hazard analysis critical control point (HACCAP)" to identify, prevent and control the food issues for best products (Laube et al., 2011; Petruzzelli et al., 2014). The knowledge of ingredients gluten content and use of the gluten-free certified ingredients by manufacturers in the food chain should be aware for manufacturing gluten-free products (Muraro et al., 2014; Mishra et al., 2020). There are many organizations for gluten-free standards like Codex Alimentarius Commission, Food and Drug Administration, European Commission, etc., and there are many third-party organizations like US Gluten-free certification organization (GFCO), celiac support organization, etc., that work with government policies for gluten-free products (Casper & Atwell, 2016). There is significant variation among the emerged regulatory framework of organizations irrespective of international and scientific collaboration on gluten issues. Effective avoidance of gluten depends upon information acquired from food labels, people having complete and precise information on substance in food (Gendel, 2012). The organizations for Gluten-free regulations require a written plan for food safety guidelines that includes processing practices, tolerant level of gluten, food label, misleading policies, recall options, etc., for the product (Crawford, 2019; Desmarchelier & Szabo, 2008). Codex Alimentarius international food standards are the regulatory standards from which most of the countries adopt the policies and standards for gluten-free labeling and gluten quantitative detection (Lee et al., 2014).

## 6.3 Codex Alimentarius International Food Standards

The implementation of Codex Alimentarius commission standards (CXS 118-1979) of "world health organization (WHO)" and "Food and Agricultural Organization (FAO)" for gluten-free products was developed during 1976. The law is modified in 1983 and 2015 and revised in 2008, defined the products as "Gluten-free Foods" (Arentz-Hansen et al., 2004; Bustamante et al., 2017; Jnawali et al., 2016). This standard has identified "Gluten" as the protein from oats, barley, rye, wheat, or else crossbreed type and "prolamins" as a fraction from gluten i.e. gliadin from wheat; secalin from rye; hordein from barely and also avenin from oats (Commission, 2008).

The scope of this standard is to apply for the dietary function of gluten-sensitive people, food that has been developed, manufactured, or processed. The products well-defined by the standard are: "Gluten-free foods" as gluten content should not exceed above 20 mg/kg or 10 mg/Kg gliadin in food produced from (1) Ingredients other than oats, barley, rye, wheat, or else crossbreed grain varieties. (2) ingredients i.e. oats, wheat, barley, rye, or crossbred grain varieties that are processed particularly to remove gluten; "Foods particularly processed to lower gluten content to level >20 to 100 mg/kg" as food distributed or traded to consumer-produced from ingredients rye, barley, oats, wheat or grain crossbred varieties and treated to decrease the gluten content of food to content >20 to 100 mg/kg i.e. gluten content of food should not be above 100 mg/Kg (Hager et al., 2014; Haraszi et al., 2011; Lee et al., 2014; Mattioni et al., 2016; Vassiliou, 2009). The food manufactured to avoid the contamination of gluten should follow the guidelines of good manufacturing practice (GMP). The nutritional content of original food such as vitamins, minerals should provide approximately equal dietary content even though the gluten content of food is substituted or processed (Commission, 2008).

The guidelines for gluten-content analyzing techniques in food are also stated in codex standards. The quantitative detection of gluten content in food should follow the immunologic or other technique that has the same specificity or sensitivity. These methods should be calibrated and validated with reference material and have has a detection limit of 10 mg/kg or less according to the technical standard. The relevant methods that can be used for detection are R5 Mendez Enzyme-linked immunoassay (ELISA) or DNA technique (Commission, 2008).

## 6.4 Labelling Regulations

The following codex"Gluten-free food" labeling is applied in addition to the guidelines of the "General labeling of prepacked food" and "General Standard for the Labelling of and Claims for Prepackaged Foods for Special Dietary Uses"(Commission, 2008).

1. For the products having gluten content less than 20 mg/kg, the word “Gluten-free” can be designated on the product close to the product title (Bustamante et al., 2017).
2. For the products having a gluten content of about 20–100 mg/Kg which are specially processed they are not gluten-free products. Hence, for such products, the nature of the product should be designated close to the product name.
3. The food or product which is naturally gluten-free can be indicated with the term “this food is by its nature gluten-free” on the package but the terms “special dietetic” or special dietary” or any more equivalent labels should not be indicated so that the consumer is not misled by the labeling provided.

## 6.5 Food and Drug Administration (FDA)

The Food and Drug Administration (FDA) of the united states has claimed the food label “Gluten-free” regulation in August 2014 (Jnawali et al., 2016) in which this is the first regulation in the united states for “Gluten-free foods”. The “Department of Health and Human Services (HHS)” was addressed by“Food Allergen Labeling and Consumer Protection Act of 2004 (FALCPA)” to approve and define the term “gluten-free” for practice in FDA-regulated foods labeling. The regulation was started such that “Gluten-free” labeled food provides a clear trustful standard for gluten-sensitive people to manage their health. The final rule was issued in 2013 for “Gluten-free foods” labeling and recently final was issued on August 12, 2020, for “Gluten-free” labeling of Fermented or hydrolyzed foods. The August 29, 2020, revised rule has established compliance regulations for hydrolyzed, distilled, and fermented foods in which the fermentation or distillation should do for gluten-free” ingredients only as there is no analytical method for detecting gluten content in the fermented or hydrolyzed sample. Hence, the FDA inspects the records given by the manufacture of “Gluten-free” ingredients used for claiming the label “Gluten-free”. In the united states, it was estimated that three million people are affected due to the consumption of gluten. The FDA stated that according to the regulation, the manufacture is responsible for not misleading and fulfilling all guidelines furnished by regulation (FDA, 2014).

According to FDA, the foods can be labeled “Gluten-free” for the products which comply with the following specifications (FDA, 2014):

1. The gluten content in the foods should be less than 20 ppm (Ahmad et al., 2019).
2. The food produced should not include the ingredients related to wheat, barley, rye, or grains of crossbreed.
3. The food should not include ingredient is not processed to exclude gluten content from the ingredients obtained from grains.
4. The food should not include ingredients obtained from grains that are processed to exclude gluten but have a gluten content of food above 20 ppm.

The specifications met by the food can be labeled by the terms "Gluten-Free"; "no gluten"; "without gluten"; "free of gluten" that are regulated by the FDA which is manufactured gluten-free food or naturally gluten-free food.

The FDA regulation is applied for restaurant food including other processed food cereals, bread, grain-based foods, pasta, beverages. The FDA is associated with national and local government for the gluten-free food in restaurants. Beer can bear the label "gluten-free" if it is manufactured from Gluten-free grains or ingredients treated to exclude gluten before fermentation for 21 CFR 101.91 enforcement. The food imported to the united states should meet the regulation requirement for the "Gluten-free" label (FDA, 2014).

## 6.6 European Union Standard

The European Union follows the regulations of the Codex Alimentarius Commission standards. The European Union has issued "Commission Regulation (Ec) No 41/2009" for labeling for people who are gluten-intolerant in the year 2009. The regulation guidelines are revised later in the next following years as "Regulation (EU) No 1169/2011"; "Regulation (Eu) No 1155/2013"; "Regulation (Eu) No 828/2014" and has been implemented July 20, 2016. Regulation (EU) No 1169/2011 is for the information of the presence or absence of gluten in food for gluten-sensitive people (Popping & Diaz-Amigo, 2018). The regulation is given by the EU in agreement with the acceptance of the "committee on the food chain and animal health". The scope of the standard is the same as the codex standard i.e. for gluten intolerant people, the standard is to apply for the dietary function of gluten-sensitive people, food that has been developed, manufactured, or processed. The regulatory information about "Gluten-free foods" given by food operators should not be misled or appropriately provided to the consumer (European commission).

According to the EU document, the grains or grain-derived ingredients containing gluten are reported to be Rye, barely, Wheat (Triticum species); the oats grain may exclude as it may not cause-effect to all intolerant people. The statement such as "Commission Directive 2006/125/EC" has set a regulation that the "Gluten-free" should be indicated on infant food and cereal-based food for the infant and young children below age six. The term "no-gluten" or "Gluten-free" could be labeled for the food that has a gluten content below 20 mg/Kg (Casper & Atwell, 2016), (European Commission). The regulation guidelines are given by Commission Regulation (Ec) No 41/2009 are:

1. The foods produced from ingredients from oats, rye, barley, wheat, or crossbred grain varieties and processed particularly to decrease gluten content of food should not have a level over 100 mg/kg and those foods could be indicated as "Very Low Gluten".
2. The word "Gluten-free" can be indicated on foods having content below 20 mg/kg.

3. Gluten content should not exceed 20 mg/kg for the foods having oats for gluten intolerant people, to avoid cross-contamination with other gluten-containing grains the food should be processed.
4. The label "Gluten-free" is applicable for the ingredients having gluten content below 20 mg/kg which is used as a substitute for the oats, wheat, barley, rye, or their grains of crossbreed.

The "Association of European celiac society (AOECS)" for the food having gluten content <20 mg/Kg has licensed the usage of label symbol "Cross-grain" as a quality mark (Bustamante et al., 2017).

## 6.7 Canada Regulations

The Canadian population affected by celiac disease due to gluten is estimated to be 3,40,000 people or 1% of the population. The "Health Canada" of Government of Canada has issued their first labeling regulation for "Gluten-free foods" in 1995. The "Health Canada" and "Canadian Food Inspection Agency (CFIA)" work with the industrial food manufacturers, distributors, importers, consumer associations for the effective labeling of Gluten-free, food allergens added in prepacked food for Gluten-intolerant people. The CFIA has implemented regulatory guidelines for allergen control like cross-contamination in the industry. Division 24, Food and Drug regulation (FDR) of Canada has issued a regulation for the usage of the label for "Foods of special dietary use" in which section B.24.018 issued on August 4, 2012, is regulation for the "Gluten-free food". According to the regulation, it states that "The food is banned to package, sell, package or promote if food holds any gluten protein or improved gluten protein, including a fraction of gluten protein fraction" (Health Canada, 2012).

The scientific regulation by Health Canada for designating the label "Gluten-free", the food should contain gluten content below 20 mg/Kg which is manufactured under good manufacturing practices without any cross-contamination. Health Canada 2015 has declared that people with gluten-sensitive disorders should not consume regular oats that have not been processed (Allred et al., 2017). The "list of ingredients" or "contains" label should be declared on the food if the food contained gluten is intentionally added to food even though if it is added at a small quantity (Casper & Atwell, 2016). In some cases, the term "Gluten-free" can be included if the food undergoes recognized effective processing step to remove the gluten content in cereal-derived food. The detection of the gluten content of food and processed food can be done by analytical method Enzyme-linked immunoassay (ELISA) R5 Mendez technique that is assessed as an effective method by Health Canada (2012).

The Canadian celiac association is a federally registered organization also associated with a certification program for "Gluten-free foods". This organization helps in promoting the healthiness of people suffering from diseases like celiac or other gluten- intolerance by providing information and regulatory guidelines. The gluten content threshold level is the same as issued by Health Canada i.e. less than 20 ppm (Health Canada, 2012).

## 6.8 Indian Regulation

Celiac disease and gluten- intolerant disorder has posed a public health concern and has estimated to affect 6–8 million people (Deora et al., 2014; Makharia et al., 2011). The food safety and standard, Act 2006 has started to reduce food adulteration and bring food safety in the country. The regulations by food safety and standards, Government of India has issued gazette regulation by "Food product standards and food additive" and "Food packaging and labeling" for "Gluten-free food". Regulatory labeling for gluten-free packaged food came into existence for product identification and to avoid contamination by "Food Safety and Standard Regulation (FSSR)" in 2011. The "Ministry of Health and Family Welfare (MoHFW)" in 2016 implemented new regulatory guidelines for 'Gluten-free food".

According to regulations, the term "Gluten-free foods" should be indicated for the foods having gluten content <20 mg/Kg whereas "Low-Gluten" for the foods having gluten content 20–100 ppm. The foods which are naturally gluten-free can claim the "This food by its nature gluten-free" label. The foods rice, millets, ragi, legumes, and pulses and product should not be contaminated with wheat or their respective ingredients can have the label "Gluten-free" containing gluten content below 20 mg/Kg. The "Low-gluten" for the food contains ingredients from rice, ragi, millets, oats, barley, rye, maize, wheat, legumes, and pulses which have gluten content between 20 and 100 ppm. The "Food packaging and labeling", 2016 of food safety and standard should provide a warning as "the food indicated as low gluten may poses hazard for people with celiac disease" (Dudeja et al., 2016).

## 6.9 Other Regulatory Labelling for "Gluten-Free Food"

The wheat-containing "Gluten" is one of the fourteen food allergens. In 1993, a codex committee was formed for the development of regulatory guidelines for food labeling of allergens and in 1999, a codex general standard for the labeling of pre-packed was issued. Numerous regulations, laws, labeling has been implemented by countries for the allergens control are mostly based on European commission or codex standards (Gendel, 2012; Watson, 2013). In Japan by Japanese law, allergen labeling is compulsory and cereals such as wheat, buckwheat are among them. The japan government in 2006 for labeling has recommended analytical method "Polymerase chain analysis" for the detection of Allergen in wheat, buckwheat and has regulated that allergen protein should be below 10 mg/Kg to indicate the labeling (Akiyama et al., 2011). Brazil health authority in 2002 has issued a law that all food products including beverages should indicate the gluten absence or presence from the rye, wheat, oats, barley, and their grain derivatives. The government has launched recently 2015, guidelines to help people with gluten intolerance and celiac disease (Mattioni et al., 2016). According to the Brazil regulation, the food possible to cross-contamination should be declared when the gluten ingredients are not intentionally added and the unintentional trace amounts to be avoided by the GMP and allergen management control are not adequate to avoid. It is mandatory for the warning "contain Gluten" if there is knowledge of cross-contamination by the cereal-grains or their grain derivatives (de Almeida, 2016; Pinto et al., 2020). The "Celiac disease foundation-AssociazioneItalianaCeliachia (AIC)" of Italy is a non-profit organization for celiac disease people as they gluten-intolerant people as 2,00,000 people are suffering from Gluten-intolerance by reports of Italian Health Ministry. The "National health system" of Italy provides 140 Euros/month for the "Gluten-free" products developed for Gluten-sensitivity people (Cornicelli et al., 2018) (AssociazioneItalianaCeliachia). The Argentina country regulation has lowered the threshold level to 10 ppm for the indication of "Gluten-free" label. According to regulation of chapter 17 Article 1383 of "Argentina Food code", food that is naturally free of gluten are "Gluten-free foods" and should follow GMP to avoid cross-contamination to products (Navarro et al., 2017). The Australia and New Zealand regulation are regulated in the Food standard codes of their country for "Gluten-free foods" labeling. According to their regulations, the label "Low Gluten" can be designated to foods that have gluten content less than 200 ppm while the "Gluten-free" can be given to foods prepared with no-gluten ingredients derived from even oats (<3 ppm) even that been hydrolyzed or malted (Canberra, 2017).

## 6.10 Misleading of Regulation

The countries are doing a huge effort for developing and following guidelines as per the regulations. Even though, there is a chance of misleading of the regulation intentionally or unintentionally. The gluten content more than threshold level even in less gluten quantity has a huge impact on the health of gluten-sensitive people. A study by (Verma et al., 2017) has estimated the level of contamination in certified "Gluten-free" food and naturally gluten-free products by a random collection of a sample from the Italian market which resulted that 9% of the samples have gluten content more than 20 ppm. (Halmos et al., 2018) has also experimented percentage of adulteration in Gluten-free products as part of the survey program of *Food Act 1984* by a collection of samples from food outlets in Melbourne, Australia, and has analyzed that 6% of Gluten-free samples have more than 20ppm gluten. Hence, the regulations must take active and immediate action on the food products that are violated from the rules to avoid such contamination for people. In some cases, the "Gluten-free" logo cross-grain on the package is misused by some manufacturers as they don't have the official license and place logo on the package without proper analysis of the product leading to a huge risk for the health of gluten-intolerant people as they face the problem for identification of "Gluten-free" product (Bartczak et al., 2017; Grabowicz & Czaja-Bulsa, 2019; Silvester et al., 2016).

Some organization of different countries has issued some immediate actions if the "Gluten-free" product is misleading the regulations. According to the regulatory action of FDA, the individual can report a problem to FDA regarding the mismanagement of food or label if it is against the regulation. It has issued that individuals intolerant to gluten who have suffered from illness or injury due to the consumption of the certified food should seek medical care and report the problem to FDA (2014). In some cases, withdrawal of the product is requested to be done with proper procedure of investigation of adulteration if the gluten content of product are above 20 ppm or below 100 ppm and for the gluten content above 100 mg/Kg, the product is recalled (Dudeja et al., 2016). The Canada government, Health Canada recalls the product from market and warning to the public is given. In another case of violation of the Section B.24.018 FDR if the gluten content is above 20 ppm, the Health Canada enforcement with CFIA can recall the product (Health Canada, 2012). "Food Safety and Standard Regulation (FSSR)" of India has also the provision to recall the product if it there is food adulteration.

## 6.11 Conclusion

The different Nations government and organizations have taken different approaches for the regulatory labelling of "Gluten-free" foods for the Gluten-intolerant people as it is significant problem worldwide. The respective country or organization is responsible for the development of strict guidelines for the manufacturers, retailers, and marketers to indicate the label "Gluten-free" on the package without any violation of regulations. Many countries follow the regulations of "Codex Alimentarius commission standards", Food and Drug Administration (FDA) and European union (EU) commission. These regulations have issued the label "Gluten-free" for the food containing gluten-content below 20 ppm and also indicate the absence or presence of gluten as most Gluten-intolerant people are sensitive to even a small quantity of gluten. The organizations must assist the people to avoid the food that provokes the health issue as consumption of Gluten food can cause mild to severe health problems to Gluten-sensitive people. The public should know the regulations and labels of the organization to avoid the mis-consumption of Gluten-free products. There is an increase in consumption of "Gluten-free foods" due to the health consciousness and raise in the population of people suffering from Gluten-intolerance in the world. The cost of the "Gluten-free food" is higher than normal food as the "Gluten-free food" sometimes requires special processing for the removal of gluten. The organizations should effectively take preventive measures to reduce the contamination and misleading of the sample. Even though many methods for quantitative detection of gluten content has been invented only a few methods like ELISA, DNA is recommended for analysis by many "Gluten-free" organizations. Hence, more effective, rapid, and cost-effective methods for detection of samples such that detection of gluten can be done in the in-line process of industry or by the consumers while purchasing. A few countries like India, the US, Ireland, Canada, Brazil, etc., have only issued regulations for gluten-intolerant people but the health issue of gluten-intolerance and celiac disease has also been in many other countries. Hence, the remaining countries should take risk management procedures for the people suffering from Gluten-intolerance.

## References


Ahmad, I., Swaroop, A., & Bagchi, D. (2019). *An overview of gluten-free foods and related disorders Nutraceutical and Functional Food Regulations in the United States and around the World* (pp. 75–85). Elsevier.

Akiyama, H., Imai, T., & Ebisawa, M. (2011). *Japan food allergen labeling regulation – History and evaluation* (Advances in food and nutrition research) (Vol. 62, pp. 139–171). Elsevier.

Akobeng, A. (2008). The evidence base for interventions used to maintain remission in Crohn's disease. *Alimentary Pharmacology & Therapeutics, 27*(1), 11–18.

Alimentarius, C. (2003). *Draft revised standards for glutenfree foods*. Report of 25th session of the Codex Committee on Nutrition and Foods for Special Dietary Uses. November.

Allred, L. K., Kupper, C., Iverson, G., Perry, T. B., Smith, S., & Stephen, R. (2017). Definition of the "Purity Protocol" for producing gluten-free oats. *Cereal Chemistry, 94*(3), 377–379.

Arentz-Hansen, H., Fleckenstein, B., Molberg, Ø., Scott, H., Koning, F., Jung, G., ... Sollid, L. M. (2004). The molecular basis for oat intolerance in patients with celiac disease. *PLoS Medicine, 1*(1), e1.

Association of European celiac society (AOECS). https://www.aoecs.org/aoecs-gluten-free-standard

Astley, S. (2019). *Health claims and food labelling* (Vol. 22). Royal Society of Chemistry.

Bartczak, A., Zhang, J., Oyedele Adeyi, A. A., Grant, D., Gorczynski, R., Selzner, N., ... Levy, G. A. (2017). Overexpression of fibrinogen-like protein 2 protects against T cell-induced colitis. *World Journal of Gastroenterology, 23*(15), 2673.

Bustamante, M. Á., Fernández-Gil, M. P., Churruca, I., Miranda, J., Lasa, A., Navarro, V., & Simón, E. (2017). Evolution of gluten content in cereal-based gluten-free products: An overview from 1998 to 2016. *Nutrients, 9*(1), 21.

Canberra, A. (2017). Australia New Zealand Food Standards Code: Schedule 5—Nutrient profiling scoring method.[Google Scholar].

Casper, J. L., & Atwell, W. A. (2016). *Gluten-free baked products*. Elsevier.

Celiac disease foundation-AssociazioneItalianaCeliachia (AIC). https://celiac.org/eat-gluten-free/gf-services/italian-celiac-association/#:~:text=The%20mission%20of%20AIC%20(Associazione,in%20a%20calm%20and%20aware

Commission, C. A. (2008). *Codex standard for foods for special dietary use for persons intolerant to gluten*. http://www.codexalimentarius.net/download/standards/291/cxs_118e.pdf

Cornicelli, M., Saba, M., Machello, N., Silano, M., & Neuhold, S. (2018). Nutritional composition of gluten-free food versus regular food sold in the Italian market. *Digestive and Liver Disease, 50*(12), 1305–1308.

Crawford, W. M. (2019). *The Food Safety Modernization Act (FSMA)'s role in the safety of functional foods and their ingredients Nutraceutical and Functional Food Regulations in the United States and around the World* (pp. 61–74). Elsevier.

de Almeida, M. F. L. (2016). *Monitoramento e avaliação da regulamentação sobre rotulagem de alimentos alergênicos no Brasil: proposição de indicadores e métricas*. PUC-Rio.

Deora, N., Deswal, A., Dwivedi, M., & Mishra, H. (2014). Prevalence of coeliac disease in India: A mini review. *International Journal of Latest Research Science Technology, 3*(10), 58–60.

Desmarchelier, P. M., & Szabo, E. A. (2008). Innovation, food safety and regulation. *Innovations, 10*(1), 121–131.

Dudeja, P., Dudeja, A., Singh, G., & Mukherji, S. (2016). Regulatory framework for "gluten-free" foods in India: Magic bullet for celiac disease patients. *Medical Journal of Dr. DY Patil University, 9*(6), 680.

Duttaroy, A. K. (2019). *Regulation of functional foods in European Union: Assessment of health claim by the European Food Safety Authority Nutraceutical and Functional Food Regulations in the United States and around the World* (pp. 267–276). Elsevier.

European Commision. http://data.europa.eu/eli/reg/2009/41/oj

Gendel, S. M. (2012). Comparison of international food allergen labeling regulations. *Regulatory Toxicology and Pharmacology, 63*(2), 279–285.

Grabowicz, A., & Czaja-Bulsa, G. (2019). Misleading labelling of gluten-free products in the light of EU regulations: Time for a change? *Journal of Consumer Protection and Food Safety, 14*(1), 93–95.

Hager, A.-S., Taylor, J. P., Waters, D. M., & Arendt, E. K. (2014). Gluten free beer – A review. *Trends in Food Science & Technology, 36*(1), 44–54.

Halmos, E. P., Di Bella, C. A., Webster, R., Deng, M., & Tye-Din, J. A. (2018). Gluten in "gluten-free" food from food outlets in Melbourne: A cross-sectional study. *The Medical Journal of Australia, 209*, 42–43.

Haraszi, R., Chassaigne, H., Maquet, A., & Ulberth, F. (2011). Analytical methods for detection of gluten in food – Method developments in support of food labeling legislation. *Journal of AOAC International, 94*(4), 1006–1025.

Health Canada's Position on Gluten-free food. https://www.canada.ca/content/dam/hc-sc/migration/hc-sc/fn-an/alt_formats/pdf/securit/allerg/cel-coe/gluten-position-eng.pdf

Jerome, R. E., Singh, S. K., & Dwivedi, M. (2019). Process analytical technology for bakery industry: A review. *Journal of Food Process Engineering, 42*(5), e13143.

Jnawali, P., Kumar, V., & Tanwar, B. (2016). Celiac disease: Overview and considerations for development of gluten-free foods. *Food Science and Human Wellness, 5*(4), 169–176.

Laube, T., Kergaravat, S., Fabiano, S., Hernández, S., Alegret, S., & Pividori, M. (2011). Magneto immunosensor for gliadin detection in gluten-free foodstuff: Towards food safety for celiac patients. *Biosensors and Bioelectronics, 27*(1), 46–52.

Lee, H. J., Anderson, Z., & Ryu, D. (2014). Gluten contamination in foods labeled as "gluten free" in the United States. *Journal of Food Protection, 77*(10), 1830–1833.

Madhuresh, D., Mishra, H. N., Deora, N. S., Baik, O. D., & Meda, V. (2013). A Response Surface Methodology (RSM) for optimizing the gluten free bread formulation containing hydrocolloid, modified starch and rice flour. *The Canadian Society for Bioengineering (CSBE)*, 13–112.

Makharia, G. K., Verma, A. K., Amarchand, R., Bhatnagar, S., Das, P., Goswami, A., … Anand, K. (2011). Prevalence of celiac disease in the northern part of India: A community based study. *Journal of Gastroenterology and Hepatology, 26*(5), 894–900.

Martini, D., Del Bo, C., & Cavaliere, A. (2019). *Current legislation in the European context: A focus on food labeling, novel foods, nutrition, and health claims Nutraceutical and Functional Food Regulations in the United States and around the World* (pp. 253–265). Elsevier.

Mattioni, B., Scheuer, P. M., Antunes, A. L., Paulino, N., & de Francisco, A. (2016). Compliance with gluten-free labelling regulation in the Brazilian food industry. *Cereal Chemistry, 93*(5), 518–522.

Mishra, N., Tripathi, R., & Dwivedi, M. (2020). Development and characterization of antioxidant rich wheatgrass cupcake. *Carpathian Journal of Food Science & Technology, 12*(3).

Muraro, A., Hoffmann-Sommergruber, K., Holzhauser, T., Poulsen, L. K., Gowland, M., Akdis, C., … Schnadt, S. (2014). EAACI Food Allergy and Anaphylaxis Guidelines. Protecting consumers with food allergies: Understanding food consumption, meeting regulations and identifying unmet needs. *Allergy, 69*(11), 1464–1472.

Navarro, V., del Pilar Fernández-Gil, M., Simón, E., & Bustamante, M. Á. (2017). *Gluten: General aspects and international regulations for products for celiac people nutritional and analytical approaches of gluten-free diet in celiac disease* (pp. 15–27). Springer.

Petruzzelli, A., Foglini, M., Paolini, F., Framboas, M., Serena Altissimi, M., Naceur Haouet, M., … Cenci, T. (2014). Evaluation of the quality of foods for special diets produced in a school catering facility within a HACCP-based approach: A case study. *International Journal of Environmental Health Research, 24*(1), 73–81.

Pinto, C. A., de Souza, B. R., Peixoto, J. d. S. G., & Ishizawa, T. A. (2020). Rotulagem para alergênicos: uma avaliação dos rótulos comercializados com presença ou ausência de glúten e seus riscos inerentes a saúde dos celíacos no Brasil. *Research, Society and Development, 9*(6), 4.

Popping, B., & Diaz-Amigo, C. (2018). European regulations for labeling requirements for food allergens and substances causing intolerances: History and future. *Journal of AOAC International, 101*(1), 2–7.

Rifna, E. J., Singh, S. K., Chakraborty, S., & Dwivedi, M. (2019). Effect of thermal and non-thermal techniques for microbial safety in food powder: Recent advances. *Food Research International, 126*, 108654.

Rosell, C. M., Barro, F., Sousa, C., & Mena, M. C. (2014). Cereals for developing gluten-free products and analytical tools for gluten detection. *Journal of Cereal Science, 59*(3), 354–364.

Rostami, K., Bold, J., Parr, A., & Johnson, M. W. (2017). *Gluten-free diet indications, safety, quality, labels, and challenges*. Multidisciplinary Digital Publishing Institute.

Sainsbury, K., Halmos, E. P., Knowles, S., Mullan, B., & Tye-Din, J. A. (2018). Maintenance of a gluten free diet in coeliac disease: The roles of self-regulation, habit, psychological resources, motivation, support, and goal priority. *Appetite, 125*, 356–366.

Sharma, G. M., Pereira, M., & Williams, K. M. (2015). Gluten detection in foods available in the United States – A market survey. *Food Chemistry, 169*, 120–126.

Silvester, J. A., Weiten, D., Graff, L. A., Walker, J. R., & Duerksen, D. R. (2016). Is it gluten-free? Relationship between self-reported gluten-free diet adherence and knowledge of gluten content of foods. *Nutrition, 32*(7–8), 777–783.

Thompson, T., & Méndez, E. (2008). Commercial assays to assess gluten content of gluten-free foods: Why they are not created equal. *Journal of the American Dietetic Association, 108*(10), 1682–1687.

Tovoli, F., Masi, C., Guidetti, E., Negrini, G., Paterini, P., & Bolondi, L. (2015). Clinical and diagnostic aspects of gluten related disorders. *World Journal of Clinical Cases: WJCC, 3*(3), 275.

Tripathi, R., Sharma, D., Dwivedi, M., Rizvi, S. I., & Mishra, N. (2017). Wheatgrass incorporation as a viable strategy to enhance nutritional quality of an edible formulation. *Annals of Phytomedicine, 6*(1), 68–75.

U.S Food and Drug Administration (FDA) for Gluten and Food Labelling. https://www.fda.gov/food/nutrition-education-resources-materials/gluten-and-food-labeling#:~:text=The%20rule%20specifies%2C%20among%20other,using%20scientifically%20validated%20analytical%20methods

Vassiliou, A. (2009). Commission regulation (EC) No. 41/2009 of 20 January 2009 concerning the composition and labelling of foodstuffs suitable for people intolerant to gluten. *Official Journal of the European Union, 21*, L16.

Verma, A. K., Gatti, S., Galeazzi, T., Monachesi, C., Padella, L., Baldo, G. D., … Catassi, C. (2017). Gluten contamination in naturally or labeled gluten-free products marketed in Italy. *Nutrients, 9*(2), 115.

Watson, R. (2013). Managing allergens from a food retailer perspective including an update on allergen labelling regulation. *Nutrition Bulletin, 38*(4), 405–409.

# Chapter 7
# Gluten Detection in Foods

**Mohona Munshi and Saptashish Deb**

**Abstract** Celiac disease is one of the most continuous perpetual food prejudices prompted by ingestion of gluten protein from various foods such as wheat, barley, rye, oats, etc. The accessibility of the various analytical methods is extremely important to determine the presence of gluten in the food matrix to guarantee the affluence of gluten delicate people. Along with it in accordance to Codex, foods having below 20 mg gluten/kg can only be considered under a gluten-free label. This also sets standards for the analytical methods in gluten detection. The present chapter deals with the chemical constituents, toxicity, and tolerance limit of gluten protein along with the importance of gluten-free foods, gluten labeling, and risk management. However, the main objective is to discuss the various gluten detection methods, their applicability, challenges, and influencing factors. Removal of gluten is necessary to increase the availability of gluten-free food products which greatly depends on the selection and standard of the detection method. Thus, to ensure the consumption of gluten-free food products the detection method of gluten must be monitored carefully.

**Keywords** Analytical methods · Celiac disease · Consumer · Gluten detection techniques

M. Munshi
Department of Food Engineering and Technology, Sant Longowal Institute of Engineering & Technology, Sangrur, Punjab, India

Department of Chemical Engineering, Vignan Foundation for Science, Technology, and Research, Vadlamudi, Guntur, Andhra Pradesh, India

S. Deb (✉)
Department of Food Engineering and Technology, Sant Longowal Institute of Engineering & Technology, Sangrur, Punjab, India

Center for Rural Development and Technology, Indian Institute of Technology Delhi, New Delhi, India



N. Singh Deora et al. (eds.), *Challenges and Potential Solutions in Gluten Free Product Development*, Food Engineering Series,
https://doi.org/10.1007/978-3-030-88697-4_7

## 7.1 Introduction

Coeliac disease is a typical issue among populaces (from 0.1% to >1.6%) of many nations. This long-lasting ailment is activated by a protein fraction called gluten available in the wheat, barley, rye, and oats instigating distinctive harm to the small bowel mucosa (Sandberg et al., 2003). Gluten is a mixture of gliadins and glutenins. Wheat, barley, rye comprise 8–15% of protein, and gluten is the main protein among others (Rosell et al., 2014). Though oats were considered harmful for coeliac patients, but clinical research and *in vitro* trials have been demonstrated which manifested oats to be endured by the vast majority of the coeliac and other aged patients (Sandberg et al., 2003).

Gluten plays a significant role in food preparation because of its inimitable physicochemical and functional characteristics. Gluten is usually used during the making of dough as a thickener to improve the texture, water binding capacity, fat binding capacity, and extensibility in various foods (Sharma & Rallabhandi, 2015). However, due to the expanded consciousness among people with coeliac disease, gluten affectability, and allergic effect, the demand for gluten-free (GF) foods has increased exponentially (Sharma et al., 2015), because the lone treatment for coeliac disease and other gluten-related diseases is the strict following of a lifetime consumption of GF food. Food items marked as GF food must comprehend below 20 mg gluten/kg (Lee et al., 2014). According to the US Food and Drug Administration (FDA), gluten-free food items should not contain any gluten-containing grain or any ingredients derived from gluten-containing grains without decreasing the gluten content below 20 mg gluten/kg (Lee et al., 2014). But, universally there is no such agreement regarding the limit of gluten content in gluten-free foods, such as according to the Spanish Federation of CD Patients Associations gluten-free foods should not contain more than 10 mg/kg of gluten for consumer safety. So far a general guideline is available regarding the analytical technique issued by Codex Standard that gluten should be separated from the food items using 60% ethanol followed by quantification using gluten detection method (Wieser, 2008). However, there are particularly some sensitive patients who can face symptomatology after the digestion of an insignificant quantity of gluten that makes all the minimum limiting points inadequate, set by different authorities throughout the world (Amaya-González et al., 2015).

Nevertheless, gluten can also accidentally get acquainted with food because of cross-contamination of inalienably gluten-free grain with gluten-containing grains during harvesting, transportation, or storage. Simultaneously, cross-contamination may happen during processing and manufacturing, when utilizing the same equipment and machinery. The inadvertent presence of gluten in GF items might be acceptable for maximum consumers yet can bring extreme reactions in gluten-sensitive people (Sharma et al., 2015). Therefore, it is essential to evaluate the gluten content in food items for labeling and marking consistency and customer safety. However, various researches demonstrated that it is hard to measure gluten for monitoring and quality control purposes, particularly after processing and cooking

(Skerritt & Hill, 1990). Gluten contains a few intermolecular disulfide bridges, those of which get opened during the time of heating, and then the intermolecular cross-linking appears after ensuing cooling. The subsequent macromolecular aggregates oppose ethanolic extraction and consequently get away from detection (Mothes & Stern, 2003).

Different investigative techniques are utilized to measure the gluten traces for assuring food safety. These are like; enzyme-linked immunosorbent assay (ELISA), polymerase chain reaction (PCR), lateral flow devices (LFD), matrix-assisted laser desorption/ionization-time of flight mass spectrometry method (MALDI-TOF), liquid chromatography-mass spectrometry (LC-MS/MS), sodium dodecyl sulfate-polyacrylamide gel electrophoresis (SDS PAGE), antibody analysis, size exclusion chromatography, biosensors, etc. All these methods contrast extensively as for their cost, affectability, and explicitness. All the measuring techniques apart from PCR distinguish either entire or part of the gluten protein, whereas PCR recognizes the gene encoding gluten protein (Sharma & Rallabhandi, 2015). Though, all the current gluten detection techniques do not relate to the least necessities of affectability, selectivity, and exactness of a gluten reference. Till today, it has not been conceivable to develop the finest and precise gluten detection techniques, and scientists from all over the world continuously searching for an accurate gluten detection method for more than the last two decades (Wieser, 2008).

Keeping in view the above points the main aim of this chapter is to elaborately explain the various available gluten detection techniques, along with their precision, selectivity, and sensitivity as well as extraction procedure, detection method, and compatibility. Simultaneously, for a clear understanding of the readers, an effort has been made to briefly discuss about the chemical constituents, poisonousness and threshold limit of gluten, the importance of gluten-free foods, and gluten labeling and risk management. However, the main focus was to present the state-of-the-art in the area of importance of gluten detection and is intended to be used as a specialized reference chapter by industry people and academicians.

## 7.2 Chemical Constituents of Gluten Proteins

Gluten is a storage protein. According to Osborne's 1924 classification, based on solubility there are four types of storage proteins. They are albumins, globulins, gliadins or prolamins, and glutenins (Shewry, 2009). Albumin is water-soluble, globulin is salt soluble, gliadin or prolamins is aqueous alcohol soluble, and glutenin is either acid or alkali-soluble. Wheat, rye, and barley are considered the rich and major source of gluten, and the starchy endosperm of the seed of these grains is the storage house of gluten. Gluten acts as a source of carbon, nitrogen, and sulfur at the time of seed germination, apart from this there is no other known biological role of gluten proteins (Shewry & Halford, 2002). Gluten alludes to a heterologous group. Gluten is a high mixture of proteins and made out of gliadin and glutenin protein fractions. Gliadin comprises 30–40% of total proteins, whereas glutenins comprise around 45% of total proteins. However, the content of soluble and insoluble glutenin varies from 9.5% to 14.9% and 24.6% to 26.7%, respectively (Žilić, 2013). The name of gliadin varies from source to sources such as gliadin for wheat, secalin for rye, hordein for barley, and avenin for oats. However, in general, gliadin is known as prolamin (Sharma et al., 2014). Along with gliadins and glutenins, gluten contains hundreds of other protein fractions, which are basically monomers, polymers, or internally connected disulfide bonds (Žilić, 2013).

The structure and functionality of gluten mainly depend on cystenine. Cystenine is one of the minor amino acids present (≈2%) in gluten protein (Wieser, 2003). Based on structure gliadin and glutenin are totally different. Gliadins are monomeric in the structure having 30–70 kDa molecular masses (Lee et al., 2014) and glutenins are large polymeric. Subdivisions of gliadins are α- (30–41 kDa), γ- (42–51 kDa), and ω- (52–74 kDa) gliadins, whereas the content of ω-gliadins is extremely inconstant in various wheat cultivars and mainly depends on fertilization. Molecular masses of various ω- gliadins (i.e. ω5-, ω1-, and ω2-gliadins) varies between 30 and 74 kDa (Mothes & Stern, 2003; Žilić, 2013). On the other hand subdivisions of glutenins are low molecular weight (LMW) glutenins and high molecular weight (HMW) glutenins. High molecular weight glutenin is again divided into x-type high molecular weight glutenins (75–120 kDa) and y-type high molecular weight glutenins (75–120 kDa). In general, the size of low and high molecular weight glutenins varies from ~270–330 residues long and ~ 650–800 residues long, respectively (Haraszi et al., 2011). High molecular weight glutenin (20–30% of total proteins) and low molecular weight glutenin further are divided into four different subgroups. The subgroups are A, B, C, and D. Subgroup A comes under high molecular weight glutenin, and subgroups B, C, and D come under low molecular weight glutenin. The molecular weight of subgroups B, C, and D varies in between 42 to 51 kDa, 30 to 41 kDa, and 52 to 74 kDa, respectively (Žilić, 2013). Subgroups B and C are the main constituents of low molecular weight glutenin and comprise around 60% of total low molecular weight glutenin. Despite all these various subunits and their size ranges, the actual size of the largest glutenin polymers is yet to discover. Therefore, glutenin proteins are considered the largest protein molecules in nature (Wrigley, 1996). This size variation of glutenin proteins affects the quality of the end products (dough) such as viscoelasticity and strength. Monomeric gliadins and polymeric glutenins are the rich sources of asparagine, glutamine, and arginine but the concentration of essential amino acids such as lysine, tryptophan, and methionine are very less (Shewry, 2007). However, better food preparedness and baking nature of bread mainly depend on a higher proportion of monomeric and polymeric gluten protein (Park et al., 2006).

## 7.3 Perniciousness of Gluten Proteins Around the World

Celiac disease is a typical immune system provocative disease with both hereditary and ecological segments; it is basically a cell-intervened immune system ailment that is known to be evoked by gluten mostly in hereditarily inclined people. Celiac disease is one of the chronic diseases, being evaluated to influence around 1% of the world population, in spite of the fact that this number could be thought little of attributable to challenges in diagnosing this immune system issue (Denham & Hill, 2013). The indications of these issues may fluctuate, contingent upon singular affectability and illness seriousness (Sharma et al., 2015). Symptoms of celiac disease may include diarrhea, iron deficiency, nausea, loss of weight, stunting, abdominal pain, etc. (Lee et al., 2014). Apart from celiac disease, 0.2–0.5% of the world population faced other food allergy-related issues due to gluten protein (Zuidmeer et al., 2008).

Research showed that out of two main fractions of gluten i.e. gliadin and glutenin, gliadin fraction is more toxic. All the gliadin fractions (α-, γ-, and ω- gliadins) were found toxic through *in-vitro* and *in-vivo* experiments, whereas proof regarding the toxicity effect of glutenin is still inadequate. However, research showed that high molecular weight glutenin intensifies celiac disease similar to gliadin (Wieser, 2003). Previously celiac disease was viewed as an uncommon issue, influencing mostly the people of European countries, and in earlier days diagnosis of celiac disease purely depended on small intestinal biopsy. But the development of various specific testing tools such as antigliadin, antiendomysium, and anti-transglutaminase antibodies with the passage of time started showing that celiac disease is more vulnerable than it was believed. A colossal number of studies have as of late indicate that celiac disease is one of the long-lasting disease influencing humanity everywhere throughout the world, and celiac disease is not only affecting people in Europe and developed countries even people from developing countries such as India, North Africa, and the Middle East are also affected by this lifelong disorder (Catassi & Yachha, 2016). Celiac disease may remain in the human body in silent form and only can be detected by serological testing (Catassi et al., 1996) indicating that many celiac patients are being undiagnosed without serological screening. Celiac disease is increasing and becoming a common disease in developed and developing countries and interestingly the exact reason behind such increment is still unknown. Previously in developing countries, celiac disease was underestimated due to lack of awareness, infrastructure for diagnosis, and a misconception that celiac disease does not exist in developing countries. But the development of medicinal science day by day is showing that celiac disease is increasing in developing countries also. An increasing trend towards adopting the western food habit may be one of the reasons behind the increasing graph of celiac disease in developing countries, though this is not the only reason. To control celiac disease, developed countries have started giving more emphasis on gluten-free diets from long back, but treatment and control in developing countries are becoming more difficult as the diagnosis is a difficult task. So it is always better to give more emphasis on local gluten-free cereal products such as millet, manioc, rice, etc. (Catassi & Yachha, 2016).

Sporadically celiac disease has become a serious ailment, portrayed by various chronic diseases, and in many developing nations, it may increase in the coming days. To fight this disease and to save the maximum number of people diagnosis is very essential. Therefore, there is a need to increase the rate of diagnosis mostly in developing countries as well as the awareness regarding the severity of the disease. Simultaneously, the development of a cost-effective diagnosis technique could altogether diminish the grimness and mortality rate related to the untreated disease.

## 7.4 Gluten-Free Foods and Its Importance

Gluten-free diets originated back in 1941 to protect people from celiac disease. Till now ailment from various health problems such as obesity, bowel syndrome, and inflammation along with its number of characteristic factors such as organoleptic properties, sensorial attributes availability of variety, and various nutritional implications of a gluten-free diet has significantly raised the sales and consumption of these products (Xhakollari et al., 2018; Sharma et al., 2014). It has also been observed that the nutritional inference of a gluten-free diet is always better than other diets because it helps in relieving symptoms like gastrointestinal healing; gluten-related disorders such as osteoporosis in females, metabolic bone disease, etc. As gluten-free items are high in fat, sugar, sodium compared to other products so it produces a very good mouthfeel.

Celiac disease is a multi-system disorder and a gluten-free diet is the only treatment for this disorder, therefore, it is very significant to understand the importance of a gluten-free diet (Niewinski, 2008). Analytical methods used for gluten detection should have sensitivity so that it can provide information about daily gluten amounts that can be tolerated by the patients. Demands of gluten-free diets are not only increasing due to celiac patients but also consumers think a gluten-free diet is the healthier option. However, gluten-free products are mostly consumed by celiac disease people. Food products having the profile of free from gluten are products that do not exceed 20 mg/kg of gluten in the food items. Nonetheless, non-celiac people also prefer the gluten-free diet because the other members of the family having a celiac patient can inherit the disease. So, to stop the symptoms, from starting itself gluten-free diet is preferred. It has been found that a gluten-free diet contains devoid of many micronutrients such as iron, niacin, magnesium and they are not typically fortified while processing gluten-free products so it is always preferred to have products which are naturally gluten-free (Newberry et al., 2017). So, we can say that the current therapy of celiac disease is a gluten-free diet and consumers can prefer the gluten-free diet for the ailment of various diseases. Further consumer knowledge, attitude towards gluten-free diet has preferably increased. So, the food industry and researchers should notably think about the increase of gluten-free products or products with low gluten and also the development of gluten-detection methods.

## 7.5 Gluten Labeling and Risk Management

Customers are dependent on the labeling of food products at the time of buying with some particular specifications and a list of ingredients. Labeling is highly important for information regarding the nutritional values of the product as well as regarding the risk factors of the product for various groups of people. Labeling about the gluten content is extremely important for gluten intolerable people or celiac patients. To manage the risk factor of gluten, various regulatory bodies have made it mandatory to label the quantity of gluten on gluten-free foods and it is extremely important to quantify and label the gluten content from a consumer safety point of view. Because the various level of gluten found responsible for a different kind of toxicity for some people. 20 ppm of gluten mostly considered the threshold limit for gluten-free foods (Sharma et al., 2015). Though worldwide this limit is not a fixed limit as already discussed earlier in this chapter. In earlier days detection of gluten and labeling of gluten content was specified for specific food products or only for single ingredient-based food products. Therefore, information regarding the gluten content of complex food products is very limited to date, which is a serious threat to risk management practice (Sharma et al., 2015). Unavailability of efficient and proper detection methods, chances of cross-contamination, and finally negligence towards proper labeling basically increases the risk factors for gluten intolerance people. A survey conducted by Sharma et al. (2015) in the United States showed 1.1% of gluten-free labeled food products contained more than 20 ppm gluten. As there is no specific treatment for celiac disease or gluten intolerance and avoidance of gluten-containing food is the only way out for celiac patients and gluten intolerance people, therefore, availability and accuracy of evidence on labeling is the only option to get the information regarding the food product that they want to buy. To make the labeled information understandable by all the consumers irrespective of educational qualification, there is a need to introduce simple, hassle-free technique using local languages to explain the information regarding the ingredients of the food products and the risk factors such as gluten label for various kinds of consumers has become extremely necessary from present situation point of view. However, the accessibility and clarity of labeled data regarding the availability of allergic food ingredients and their risk factors that are purposefully utilized and added in a food item have been improved essentially over the last two decades (Hattersley et al., 2014). But more strictness and monitoring are required on guidelines and implementations of gluten content labeling in gluten-free foods especially in developing countries to minimize the risk of celiac and other gluten allergic diseases for the betterment of the future generation.

## 7.6 Various Gluten Detection Techniques

Many analytical methods have been developed to trace gluten for making it safe for human consumption and which follows the regulations laid by Codex Alimentarius. Gluten detection is similar to food allergen detecting techniques but the critical point for gluten detection is the follow-up of the correct standard. The standards are representative of the gluten proteins to be analyzed and therefore, there should be standard methods which can be followed for the gluten detection in the food samples. Therefore various detection methods are employed depending on different criteria and requirements. Some of the most commonly used gluten detection techniques have been discussed in this chapter along with their working principle and application.

### 7.6.1 *Immunological Methods*

These techniques have grown with years since 1985 for the determination of prolamin protein and have also faced many provocations for accurately measuring the prolamin content in food. An ideal antibody use for the detection of gluten should act as a reliable indicator for the detection of toxic cereal proteins. Immunological methods are widely used but the result shows very poor reproducibility because of using different standards and protocols. Use of ELISA (Enzyme-Linked Immuno Sorbent Assay) is recommended by the Codex for gluten analysis in a frequent manner. New techniques such as immune sensors, immune blotting or immune magnetic beads, and multiplex immunoassay are also being developed which is discussed in the latter part of this chapter.

### 7.6.2 *Antibody Analysis*

A healthy system of our body is protected against foreign materials or allergens or pathogens continuously with the help of a highly sophisticated and complex network known as antibodies. It includes both cellular and hormonal responses which can be changed to produce a specific antibody for a specific target choice. These are proteins having types of isotopes such as (IgA, IgH, IgG, IgE, IgD) and is used as immunoassays. These antibodies with which they bind are known as epitope with the amino acids. So, it is important to identify the type of epitopes because the analysis of the protein of the processed food depends on this feature (Harlow & Lane, 1988, 1998; Howard & Betell, 2000).

There are basically two types of antibodies namely; monoclonal antibodies and polyclonal antibodies. Monoclonal antibodies are produced by injecting the target molecule in the immune system called immunization, in which the antibodies are in an *in vivo* system. Various practices are done to slow the release of the antigen and to boost the immune system for producing antibodies with the help of adjuvant which is mixed with the antigen while injecting. Selecting animal species to produce antibodies depends on whether to produce polyclonal or monoclonal antibodies. For example: mice are used to produce monoclonal and rabbits and sheep are used to produce polyclonal antibodies. Every animal is given doses in micrograms in the range of 2–4 weeks intervals for proper immunization. After immunization, the final sera-containing antibody is removed by the removal of the blood cells and fibrinogen. Polyclonal antibodies are complex mixtures of antibodies with different specificities. Whereas, monoclonal is with single specificity and these are if required purified and used for further immunoassay (Diaz-Amigo, 2010).

Monoclonal antibodies are mostly used over polyclonal as it has several advantages over polyclonal antibodies, they have an unlimited supply of consistent affinity and specificity. Cross-reactions with prolamins do not occur (Skerritt & Hill, 1990). Skerritt et al. (1985) studied the cereal samples such as bread wheat which was prepared for the assessment and monoclonal antibodies production. The strips were developed using techniques and were incubated which resulted in the appearance of a blue-purple spot that showed antibody binding to cereal protein. Different extractants and antibodies were used for the study and the monoclonal antibodies used for the study were later conjugated and purified. This showed that the use of direct antibody conjugates gave suitable results for detecting gluten present in foods. This way various antibodies were raised against several prolamin zepitopes. However, these depend on the antibody specificity and can be obtained using the SDS page and immunoblotting which also depends on the choice of food extractant (Skerritt et al., 1985).

In monoclonal antibody, the antibodies are produced using the hybridoma techniques, and the cereal products such as wheat, barley, oats, etc. are coated over plates following ELISA protocol and then the antibodies of interest are added and measured according to their binding to the food samples (Zhang et al., 2019). The development of hybridoma techniques has made easy detection of gluten in the foodstuffs in both processed and unprocessed foods (Skerritt & Hill, 1990). It has been seen that $PN_3$ antibodies recognize distinctly the α-gliadin and monoclonal 401.21 which reacts with HMW glutenins. It is not possible to separate glutenin and gliadin but the glutenin content possesses celiac toxicity. So, recognition of glutenin is necessary. So this can be suggested that various antibodies detect different subtypes of gluten protein to a different degree. So, gluten detection depends on the antibody and the used reference material.

### *7.6.3 ELISA Techniques*

ELISA is an immunochemical analysis that has the maximum application as a determination method in food. ELISA is sensitive to gluten in the range of mg/kg and because of its ease of usage and rapid results facilitates the manufacturers to use

ELISA onsite. These give quantitative results and are used for routine check-ups of foodstuffs. ELISA detects epitopes within the group of prolamins and uses a correction factor to estimate the gluten content. According to Codex, correction factor 2 is applied (Slot et al., 2016). Several ELISA has been developed for extraction, detection, and quantification of the food allergen. But it has been found that in most cases potential cross-reactivities with the source antisera have led to low significance (Besler, 2001).

ELISA is based on the principle of detecting antibodies that are covalently linked with an enzyme such as horsedish peroxidise, alkaline phosphatase which generates a colored chemiluminescent or fluorescent product (Scherf & Poms, 2016). ELISA serves as a dominant protocol for most of the prepared commercial kits used for food allergens determination in raw and processed foods and beverages. It is usually done using a microtiter plate and spectrophotometer and other standard laboratory equipment but the testing is expensive and often leads to delay in acquiring results (Weng et al., 2016). The common ELISA employs antibodies such as R5 and G12 monoclonal antibodies and MorinangaMIoBs as polyclonal antibodies against gluten proteins. The Skerrit antibody rose against the gliadin and also detects HMW glutelins. R5 antibody is raised against rye secalin and binds to QQPFP, QQQPP, LQPFP and QLPFP epitopes in α/β-ω, γ- gliadins (Panda & Garber, 2019). ELISA is based on different antibodies raised against different prolamins sequences and fractions. This assay can detect food regardless of the type of food (Rosell et al., 2014). Chemical modifications of gluten can change the molecular properties which in turn, the epitopes get changed which affect ELISA detection. For example: γ irradiated gliadin has shown high immunoreactivity with ELISA (Kanerva et al., 2011). Protein hydrolysis can also partially or completely change the immunoreactive epitopes (Leszczynska et al., 2003; Sharma & Rallabhandi, 2015). The first polyclonal antibody reliable sandwich and competitive ELISA was developed by Windemann et al. (1982) against gliadin and α- gliadin. So, there are two types of ELISA used for the gluten analysis such as Sandwich ELISA and Competitive ELISA.

### *7.6.4 Sandwich ELISA*

The principle of sandwich ELISA is to capture antibodies that are immobilized on the microtiter plate or wells are coated with bovine serum albumin, tween 20, and the gluten extract or food extract. The sample aliquots contain the antigen which is to be analyzed and incubated which leads to forming of antibody-antigen complex. The detection antibodies used are enzyme-labeled which on incubation binds with the antigen. So, the antigen gets sandwiched by the antibodies. The unbound enzyme antigen is washed out and the addition of enzyme-substrate gives a color changed end product which is measured spectrophotometrically. The measurement can also be taken using the dot blot visualization (Slot et al., 2016). The absorbance measured is directly proportional to the sample antigen calculated on the basis of reference protein and a calibration curve is formed. The antigen here should have a minimum of two epitopes as they separate when they bind to both antigen and enzyme-labeled antigen. SoHence, ELISA is only suitable for large antigens. Therefore, it has been observed that in hydrolyzed products such as sourdough using sandwich ELISA gives the inappropriate result (Wieser, 2008). Sandwich ELISA is used for both unheated and heated foods for gliadin detection. However, sandwich ELISA cannot detect the small fragments of prolamins (Haas-Lauterbach et al., 2012). The AOAC approved sandwich ELISA based on ω- gliadin monoclonal antibodies. This was developed by Skeritt and Hill (1990). Upon cooking or processing the ω- gliadin fraction does not get denatured. ω- gliadin assay can be used for both thermal and non-thermal proteins. But the disadvantage of ω- gliadin ELISA is it does not accurately detect the prolamins. It has been found that this assay underestimates gliadins from durum wheat and overestimates prolamins from triticale and rye and greatly underestimates hoerdin. In addition, as it is sandwich ELISA it requires two epitopes so when a protein is hydrolyzed it breaks apart and two necessary epitopes can get lacked, and therefore analysis is not accurate.

In 2006, Codex endorsed sandwich R5 ELISA based on R5 monoclonal antibody against celiac toxic epitope QQPFP and some of the closely related sequences. The advantage to this R5 ELISA is that it detects all the wheat gliadins fractions and barley hordeins and prolamin was also detected using R5 sandwich ELISA. So, R5 Sandwich ELISA is best suited for detecting gluten contamination only by intact proteins instead of hydrolyzed proteins (Thompson & Méndez, 2008). But one criticism found by Kanerva and Sontag-Strohm (2006) that it overestimates barley hordein. This monoclonal immunoassay is used to recognize the toxic pentapeptide epitope which is conserved in different cereal varieties. The epitope is used for detecting gliadin in cereals. Heat treatment does not cause any change in the epitope due to its linear and short structure which permits the measurements of gluten quantitatively even in foods that are cooked by the use of R5 antibody (Immer & Haas-Lauterbach, 2010).

Sandwich ELISA is the most used method where the antibody gets adsorbed into the solid phase and then it binds itself with the free gluten antigen. Following the antibody-antigen reaction, another antibody labeled enzyme is produced from the other species of the same specific antigen which is used for the determination of the captured sample protein. The more the antigen will be bounded the more will be enzyme-dependent color reaction will take place. To detect the trace amounts the ELISA sensitivity may be augmented by the use of biotinylated secondary antibodies and streptardin conjugated with peroxidize (Besler, 2001).

The most commonly used ELISA method is the sandwich R5 ELISA for gluten protein detection. This ELISA has the limit of quantification of 1.56 ppm of gliadins (Rosell et al., 2014). This R5 ELISA system was developed based on the reference of gliadin which is highly promising in terms of extraction, sensitivity, and low detection limits. It is a robust test with repeatability and reproducibility. At present this is the most used ELISA with a monoclonal antibody kit for gluten detection. The R5 ELISA monoclonal antibody was obtained by the immunization against ω-secalins (Mothes & Stern, 2003).

### 7.6.5 Competitive ELISA

Competitive ELISA can be used for small-size antigens with one epitope. This assay contains three components namely; antibody immobilized labeled antigen, a limit quantity enzyme-labeled antigen, and unlabelled antigen of the sample. When they are mixed with the labeled and unlabelled antigen they compete with the limited number of antibody binding sites. Unbound antigens are then washed off and enzyme substrate is added for the colored end product. The greater the sample antigen the faint will be the color enzyme-labeled antigen. Calibration curve and reference protein help to quantify the gluten protein (Wieser, 2008). Competitive ELISA can detect the small peptide fragment of prolamins in highly processed foods. Competitive ELISA shows the measured above the line of quantification which proves it to be an appropriate ELISA. Competitive ELISA based on R5 and G12 antibodies are used for the detection of the hydrolyzed gluten. But these cannot distinguish between various fermented foods; they target gliadin but do not properly estimate glutenins which also contain immunopathogenic sequences (Haas-Lauterbach et al., 2012).

Competitive ELISA equipped with a kit having a microtiter plate where the well of the plate is coated with gliadin as antigen and peroxide labelled R5 antibody is added and incubated for 30 min. The absorbance obtained is inversely proportional to prolamins concentration in the sample. The result of the assay is expressed in ppm, equivalents to the peptide. The relation between the prolamin fragments varies and therefore conversion factor into gliadin cannot be applied (Immer & Haas-Lauterbach, 2010). These can provide rapid results, inefficient to give sophisticated results, competitive ELISA is easy to handle and they are suitable for routine analysis than the sandwich system.

Competitive ELISA can determine gluten such as beer, starch, or syrups. A competitive R5 ELISA was used to determine the gluten of hydrolyzed foods and being successfully used. It is not compatible with all extraction methods but it can be extracted with ethanol extraction and that prolamin extraction is only possible using ethanol extraction (Thompson & Méndez, 2008). Competitive ELISA has a limit of quantification of 0.3 mg gluten/kg and reproducibility of ±3.6% (Wieser, 2008). It can extract all the native proteins but not the processed ones. The combination of R5 competitive ELISA with enzymatic digestion of prolamins provides a reliable quantitative determination for the partially hydrolyzed gluten (Haraszi et al., 2011). The first competitive ELISA was designed by Friis in 1988 with the help of polyclonal antibody against the gliadins but it gave negative results in rye and barley, so later on a pre-incubation step was included while before raising the polyclonal antibody to overcome this issue. The major problem related to this type of assay is the inappropriate quantification of the hydrolyzed gliadins (Immer & Haas-Lauterbach, 2010). According to Scherf and Poms (2016), competitive ELISA may be used for proteins that are intact and having small size antigen (gluten peptides) because it requires a single binding site. The mostly used RIDSCREEN gliadin competitive assays to determine gluten has 922 μg of the peptide (Scherf & Poms, 2016). This technique is not compatible with cocktail extraction solution, but combining it with UPEX (Universal Prolamin and GlutelinExtractant solution) leads to complete gluten analysis. This assay is highly sensitive and reproducible having high repeatability with a detection limit of 0.44 ppm of gliadin (Rosell et al., 2014). Mena et al. (2012) combined competitive R5 ELISA with UPEX and found it accurately determined gluten in all foods, including heat-treated and hydrolyzed foods. Simultaneously, it was observed that it goes well with both sandwich and competitive ELISA which proves UPEX solution combined with competitive ELISA a found a reliable way of detection (Mena et al., 2012).

### *7.6.6 Multiplex Immunoassay*

Multiplex immunoassay enables multiple allergen detection which reduced time, the complexity of measurement, labor requirement as well as improves the automation and ease of use when compared with the ELISA method (Wieser, 2008). Multiplex immunoassay has the capability to detect the deamidated gliadin and peptides together. The multiplex assay can analyzes samples for 14 different types of food allergens, having 7–8 major groups of gluten. Two different extraction methods were used to generate the analytical samples. The sensitivity and apparent limit of detection is <5 g/ml of food allergen and gluten (Cho et al., 2015). It is actually combining antibodies against the epitopes which are mostly related in a single detection method. So, the possibility of negative results decreases. Multiple gliadin antibodies such as R5 and G12 were used to detect the hydrolyzed food that is specified by these two antibodies. Multiplex immunoassay is sensitive enough to detect gluten in various fermented hydrolyzed foods which give a unique profile of apparent gluten concentration. The profile obtained using multiplex immunoassay shows two things such as the different ratio of gluten content in the fermented food sample and secondly the quantity of those proteins. This assay is capable to classify the food on the basis of type and degree of germination or gluten hydrolysis. Also, it can help in choosing the appropriate hydrolyzed gluten standard of compared digestion and similar peptidial composition. This ability may skew identification and enable or give accurate quantification of gluten and also help in possible immunopathogenicity (Panda et al., 2017).

Gomaa and Boye (2013) investigated the detection of gluten using a flow cytometer which is a type of multiplex immunoassay which helps in multiple allergen detection and found that its ease of use and automation reduce the extraction time compared to the ELISA. Pedersen et al. (2018) used multiplex immunoassay to detect the gluten and observed cross-reactivity provides multi antibody profiling and detects new analytes. However, the correlations would be only possible if the type and number of reference material are expanded. Though multiplex immunoassay development has numerous advantages over ELISA, still it is suffering obstacles when compared with ELISA kits such as antibody specificity, extraction procedures, and availability of materials for reference. Simultaneously, multiplex immunoassay required suitable antibodies for prolamin and glutelin, essential to be raised against the cereal-specific immunogenic epitope. Multiplex immunoassay requires high specificity because the cross-reactions in gluten-containing cereals are common. Also, the extraction procedure should be taken care of as the procedure should be able to extract both prolamin and glutelin peptides from the food sample due to the different solubility properties of prolamin and glutelin. An extra extraction step is required to be included to extract the relevant peptides as compared to the ELISA kits. In the immunochemical methods so far, reference material was required to be created. In the case of multiplex immunoassay also reference material for gliadin and glutelins are required to be created along with the reference of prolamin and glutelin for rye and barley. To detect the harmful epitopes using multiplex immunoassay highly specific antibodies are required to mix the epitopes and it should be modified by chemical treatments and for this reason, such type of material is very difficult to be standardized (Slot et al., 2016).

### *7.6.7 Lateral Flow Device and Dipstick Devices*

Apart from the immunochemical devices, demand for rapid and easy to use tests to detect the traces of gluten in food and cereals has increased (Immer & Haas-Lauterbach, 2010). Therefore, immuno chromatographic assays such as dipsticks and lateral flow devices (LFD's) are in use for getting rapid results (Melini & Melini, 2018). These kinds of devices provide qualitative results within a short duration of time (Scherf & Poms, 2016). However, these devices are less sensitive than other antibody assays but at the same time, they are inexpensive and required a small amount of samples for analyzing purposes (Cao et al., 2017). These devices are based on the principle of the presence of an inline fixed antibody conjugated with colored nanoparticles. Mostly sandwich-type method is used for getting the results in processed and unprocessed products. They are commercially available and are based on the R5, G12, and Skerrit monoclonal antibodies and a few on polyclonal antibodies (Melini & Melini, 2018). Where, LFDs are used for onsite testing at the manufacturing facility to control programs and perform hazard analysis and critical control point (HACCP) detection from storage of raw material to final product. Whenever a quantitative result of a single sample is required, the use of LFD's is recommendable. Basically, LFDs are rapid strip platform based test which detects the epitope of the prolamins group to estimate the gluten content like ELISA (Slot et al., 2016). On the other hand, sample extract flows by the dipsticks due to capillary action and reaches the antibody-covered zone followed by color is formed showing the presence of gluten. These devices can be used for swab tests for contaminants surface and to check the level of gluten contamination in raw and processed materials (Scherf & Poms, 2016). The sensitivity of commercially available gliadin dipstick is enough for gluten analysis (Immer & Haas-Lauterbach, 2010). These are designed in a sandwich format which gives a specific and reliable result and this sandwich type format uses two epitopes of an antigen. The LFD's and dipsticks when designed in competitive format rely on a single epitope and gives a positive response by decreasing the intensity of the test bind that means if anything which inhibits the background binding appears to give the positive response. As a result, the limit of detection values for this competitive format is always higher which means it is less sensitive. An advantage for competitive is that it has the ability to reflect celiac spreading biological activity which requires only a single immune pathogenic site, which makes competitive less likely to give negative results with fermented and hydrolyzed foods. But the disadvantage is that they have higher variances of the zero responses, but these can be optimized by optimizing the concentration of antigen immobilized on the surface and concentration of the detection antibody to concentrate on the target sample (Cao et al., 2017). Other than ELISA, these devices are always on demand because of their user-friendliness, rapidness of testing, and ability of onsite analysis, but the drawback is that they are qualitative or semi-quantitative, therefore, more researches are required to improvise their performance (Panda & Garber, 2019).

### 7.6.8 Immunosensor

An alternative to ELISA, immunosensors are devices which are combined with a biological component like an antibody or antigen and a physicochemical transducer. Immunosensors are made by using a robust piezoelectric transducer and photonic immobilization technique those are cheaper in cost. There is an interaction between antigen and antibody which gives signal generated due to physical or chemical changes and the fluorescent enzymes. Immunosensors are cost-effective, rapid miniaturized, and useful for onsite analysis (Scherf & Poms, 2016). Currently, numerous immunosensors are being developed by researchers for gluten detection in foods. A GMR (Giant Magneto Resistive) sensor array was developed to detect the gliadin contamination of gluten-free products with less limit of detection (Melini & Melini, 2018). A label-free gliadin immune sensor had been developed by changing the frequency of quartz crystal microbalance chips with gold nanoparticles on the surface (Chu et al., 2012). Another electrochemical immune sensor was developed with the facility of use and disposal. It is combined with the modified SPCE (Screen Printed Carbon Electrodes) and carbon CNFs (Carbon-based NanoFluorides) with proper immune recognition support. The CNFs and SPCE will improve the sensitivity of the immune sensor and also provide an efficient surface for the recognition of gliadins proteins in food (Marín-Barroso et al., 2019). A competitive magneto Immuno sensor was made using gliadin coupled with fossil activated magnetic beads and incubating along with peroxidase-conjugated anti-gliadin polyclonal antibodies for the detection of the gliadin (Laube et al., 2011). Combining of anti-gliadin polyclonal antibodies immobilized on electrode surface using dithiolself-assembled monolayers monoclonal antibody raised against the celiac toxic peptide, and an alkaline phosphatase-conjugated mouse antibody allowed the detection of gliadin in food samples by using differentiated pulse voltammetry (Nassef et al.,

2008). A disposable electro immunochemical with server printed carbon electrodes for detection of IgA and IgG and anti tTG (tissue Trans Glutaminase) autoantibodies is used. Screen-printed carbon electrode was modified with monohybrid structure and for simple adsorption, tTG was immobilized generating high stability immune sensor. The analytical signal was based on the anodic redissolution and enzyme-generated cyclic voltammetry and following the same method, the screen printed electrodes were modified using adsorption depositing gold nanoparticles (DGP) generating sensor surface. Real serum samples were successfully assayed and the results were cross-checked using an ELISA kit. The deposited gold nanoparticles antibody gave high specificity for celiac disease symptoms (Neves et al., 2012). The electrochemical detection can be done within 1 min and complete within half an hour which is much less than the other methodologies. Apart from all these it has proved to be user-friendly, sensitive, and gives selective determinations. The electrochemical sensors can detect the biomarkers of celiac disease. All the electrodes invented required a very low sample volume without compromising the sensitivity. It successfully operates against anti-gliadins, anti-TG, anti-DGP, antibodies in a multiplex system (Martín-Yerga & Costa-García, 2014). Commercially available immunosensors give useful onsite analysis but the only disadvantage is that they are usually qualitative in nature (Panda & Garber, 2019).

### *7.6.9 Western Blot*

Immunological western blot helps to tackle the low sensitivity system and two-dimensional gel electrophoresis methods. R5, G12 or anti cells, T-gliadin, and α-20 antibodies are used during the time of detection (Melini & Melini, 2018). This immune blot analysis uses sodium dodecyl sulphate polyacrylamide gel electrophoresis (SDS-PAGE) for fractional procedures. The western analysis gives information regarding antigen size when it is compared to the standard support. For preliminary identification purposes generally, antibodies such as R5-sand, G12 – sand, Skerrit, and monoclonal antibodies are used to detect or visualize the proteins. The only band visualized was from10 to 25 μg/ml gluten standards. The R5 and G12 antibodies detect gliadins and Skerrit detect glutenins and the monoclonal antibodies and also detect the glutenin and gliadins both. Though these results are consistent with extensive hydrolysis of the gluten still immune pathogenic peptides are found present in the food samples (Cao et al., 2017). Western blot was not found efficient enough and also less sensible for quantifying gluten in food samples. In order to solve this problem, after transferring every protein in one dimensional SDS- PAGE (where it was electronically transferred on a polyvinyl difluoride membrane) first proteins were adsorbed, and then the antibody was added to get the confirmation of the gluten content in food. Western blot can separate the proteins and can detect the gluten proteins according to their size and this generally should be used as a confirmatory test to ELISA. Therefore, the western blot is not a recommended method to quantify gluten but it can only be used for confirmation purposes

(Rosell et al., 2014). Western blot has more drawbacks than advantages such as it is less sensitive and not commercially available and requires expertise and also gives qualitative results (Panda & Garber, 2019).

## 7.7 Other Methods

Various other approaches were made for the successful detection of the toxic peptides present in food samples. The efficiency of these electrophoresis methods is dependent on the extraction methods which will probably help to explain the overlaps of the protein fraction during the analysis (Bean & Lookhart, 2000). The proper way of extraction is followed to avoid the problem and to improve the resolution of the protein profiles (DuPont et al., 2005; Wieser, 2008).

### *7.7.1 Electrophoresis Methods*

SDS-PAGE (Sodium Dodecyl Sulphate Polyacrylamide Gel Electrophoresis), A-PAGE (Acid Polyacrylamide Gel Electrophoresis), IEF (Iso Electric Focusing), and FZCE (Free Zone Capillary Electrophoresis) are the methods that come under the electrophoresis method.

SDS-PAGE depends on the separation by size and is used for qualitative characterization of all proteins of cereals due to the detection issues (Bietz & Wall, 1972; Kolster et al., 1992). The detection of prolamin content in oats and rice and the HMW-GS of wheat and barley are done by using SDS-PAGE (Payne et al., 1979; Lookhart & Wrigley, 1995). For larger polymers such as wheat glutenins, multi-stacking electrophoresis is developed for better separation purposes (Khan & Huckle, 1992).

A-PAGE method is based on the charge density of protein and is used for fingerprinting. A-PAGE is generally used for the separation of α-, ß-, γ-, and ω- gliadins for wheat, hoerdins for barley, prolamins for rice and avenins for oats, respectively (Lookhart & Wrigley, 1995; Woychik et al., 1961).

IEF depends on different isoelectric points of protein fractions and they are separated from immobilized pH gradient. This is generally used for the separation of storage proteins and along with that, it uses strong solubilizing agents such as chaotropic like urea and thiourea for protein solubility which is difficult to maintain during IEF separation (Righetti & Bosisio, 1981).

FZCE is basically working on the basis of protein difference charge density which produces analogous separation such as A-PAGE. But FZCE has various limitations and problems so it is not favorable (Werner et al., 1994; Bean & Lookhart, 2000).

### 7.7.2 Chromatographic Techniques

HPLC (High-Performance Liquid Chromatography), RP-HPLC (Reversed-Phase High-Performance Liquid Chromatography), and Ultra-High-Performance Liquid Chromatography (UHPLC) are the chromatographic techniques.

One-dimensional HPLC is used for the separation of cereal proteins. RP-HPLC is also used for the characterization of protein size, polymorphism of protein, and biochemical characteristics. RP-HPLC along with gel electrophoresis gives improved characterized of cereal protein. On the other hand, UHPLC can be coupled with medium or high-resolution detection techniques. This technique reduces the even time of the chromatography significantly (Haraszi et al., 2011).

### 7.7.3 Immunoanalytical Techniques

Double immunodiffusion technique, dot immunoblotting technique, and rocket immunoelectrophoresis (RIE) are the immunoanalytical techniques. These methods include the use of monoclonal and polyclonal antibodies. Double immunodiffusion is a method that allows differentiating between identical and non-identical sample proteins. Yman et al. (1994) showed the applicability of this technique and its ease of procedure while detecting gluten in pasta and buckwheat products. However, this technique is not very popular because of its low sensitivity, higher time consumption, and it can only be used for quantitative measurements. On the other hand dot immunoblotting is a recent method for the detection of the food allergen. In this technique, the samples are spotted over a polyester cloth which is already pre-coated with antibodies against the allergen protein and this is immune detected using a secondary antibody. This method is highly sensitive and also inexpensive (Besler, 2001). Whereas, RIE is an old method that includes an antibody with gel where sample proteins move according to so their electrophoretic ability until the antigen-antibody complex gets precipitated into gel building rockets at a constant antigen-antibody ratio. Though this method is reliable, specific, and sensitive the major problem arises during the gel preparation and immune staining procedure, and because of that RIE is rarely used for detection purposes (Besler, 2001).

### 7.7.4 Non-immunological Methods

There are various quantitative methods for prolamin detections complementing the immunological systems by using an antibody. Non-immunological includes the genomic and proteomic methods which do not include any use of an antibody. There are alternatives to confirm the results of the immunological methods especially during the time of complex food analysis.

### 7.7.5 Proteomic Based Methods

The proteomic method in the field of gluten detection complements well with other techniques for the analysis of gluten. Proteomic methods can be applied to foods already having a low level of gluten protein. Proteomic approach is based on two components, separation of protein and identifying the proteins. Proteomics is a tool to separate, characterize and identify and also relate it with various cereal qualities. Proteomics has proven itself to be effective in protein identification and due to this toxic protein identification has increased. Mass spectrometry is one of the important tools used for its high sensitivity for the identification, characterization, and quantifying the cereal protein. MALDI-TOF (Matrix-Assisted Laser Desorption/Ionization Time of Flight Mass Spectrometry Method) was the first method used for the detection of toxic prolamins followed by the LC-MS/MS (Liquid Chromatography-Mass Spectrometry) method.

### 7.7.6 MALDI-TOF

In proteomics, identifying protein by mass spectrometry method and replacing the traditional technologies like the sequencing technology and introducing soft ionization techniques such as protein mass spectrometry is well to do techniques. MALDI-TOF works based on the principle of MALDI by generating ions from solid phase samples in high vacuum by using short laser pulses. They are accelerated using the electric field into a TOF mass analyzer. The flight time is directly proportional to the mass of the analyte and therefore the analytes used can be calibrated using the known mass. MALDI can be also coupled with an ion trap and quadrupole analyzer, other than TOF. It is a more sensitive method for protein identification which also involves the enzymatic digestion of proteins. The direct identification and molecular mass determination of protein without afore separation can be achieved using MALDI-TOF-MS. Gliadin quantification in processed and unprocessed food is carried out using MALDI-TOF-MS, altogether with the screening of the toxic cereal prolamins. MALDI-TOF can easily distinguish between prolamins of different origin even if they are present in complex food matrix (Haraszi et al., 2011). MALDI-TOF-MS ionization essentially produces mono charge ions and therefore they do not require any deconvolution step. This is an emerging alternative technique which can characterize storage proteins because of its robustness and ability to ionize intact proteins and also can tolerate the contaminate present used for gluten extraction, but this is not directly hyphenated with LC (Alves et al., 2019).

MALDI-TOF is used to determine the molecular proteins. This can simultaneously measure mass proteins without any chromatographic purification within few minutes in the low picomol range. MALDI-TOF can be divided into three divisions. They are binding analyte matrix, ionization and desorption using a laser, and separation and detecting using a mass spectrometer. The sample dissolved in the solvent and mixed with an analyte put in a metal plate and dried in the vacuum. Laser light is fired, whereupon the sample gets ionized and is carried in the vapor phase. The ions are accelerated using voltage impulses and separating in accordance with their mass, charge ratios by measuring time. MALDI-TOF was the first proteomic technique used for the evaluation of intact gluten proteins in the sample. Gliadin, secalin, hoerdin, and avenins from wheat, rye, oat, and barley respectively taken from cultivars showed the characteristic profiles within the range of 20,000–40,000 $M_r$, determining the presence of a small amount of gliadin even after 2 steps of extraction. The detection limit of MALDI-TOF is always high (Dostalek et al., 2009). Camafeita and Méndez (1998) worked on MALDI-TOF/MS to detect avenins from the cereals and they found it is suitable to detect avenin from oats, which is not possible using ELISA. Altogether MALDI-TOF is a highly valuable non-immunological method for the detection and quantification of gluten in foods.

However, MALDI-TOF-MS can be used only for semi-quantitative measurements due to its insufficient sensitivity and low accuracy for high mass. These drawbacks can be overcome by using the HPLC separation detection (LC-MS/MS) method (Scherf & Poms, 2016). Therefore, TANDEM combines with HPLC is used to increase the detection power (Immer & Haas-Lauterbach, 2010). Simultaneously, this tool does not have high specificity of gliadin determination in wheat as well as MALDI-TOF-MS is an expensive tool and requires trained personnel for its operation.

### 7.7.7 LC-MS/MS

LC-MS/MS is a proteomics-based technique which is used for the routine check-up of the food samples. It has the ability to detect various species based on multiple markers with multiple points of confirmation as a result it gives fewer negative results and because of its specificity, it has the capability of differentiating between multiple peptide markers (Shefcheck & Musser, 2004; Hernando et al., 2008; Lock et al., 2010). There are basically two approaches present in LC-MS/MS namely, untargeted and targeted approaches. The untargeted approach gives a comprehensive profile of the proteome sample and the targeted approach selects the specific molecule (Saghatelian & Cravatt, 2005). One of the most important points of LC-MS/MS is that it can determine and differentiate various peptides or proteins from various products. LC-MS/MS is sensitive and versatile and allergens can easily be detected using multi-methods. Earlier for enzyme digestion purposes, trypsin was used in LC-MS/MS during gluten profiling. At present some peptide markers such as pepsin, chymotrypsin are found unique and being used as an alternative to detect wheat gliadin proteins. However, this method takes a longer digestion time and labeling chemistry is done for better characterization. LC-MS/MS uses several peptide markers with multiple M2Ms for every peptide to specify the presence of gluten in the sample. LC-MS/MS has the ability to detect gluten even in processed foods. ELISA kits have failed to detect allergens due to the processing changes and changes of the protein structure in processed food which does not allow the antibody to bind and lead to negative results, but LC-MS/MS can overcome this issue. LC-MS/MS can be linked with genomics as well as to improve the knowledge of the gene responsible for the expressive allergenic proteins. LC-MS/MS has been used as a complementary tool to confirm the results produced using ELISA and to check the immunogenic peptides (Scherf & Poms, 2016). THE current LC-MS/MS method needs a four-time dilution of the sample before putting into the system to get the result. Due to the presence of multiple markers for gluten variety which gives multiple point configurations of gluten, time requirement and chances of false result in LC-MS/MS is very less (Lock, 2014).

Liao et al. (2017) in their studies shows that LC-MS/MS can detect the hoerdin from a gluten-free beer with a high degree of identification. LC-MS/MS can combine with HPLC chromatography with a mass spectrometer where HPLC can digest the peptides on the basis of properties such as size change, hydrophobicity, etc. and also removes the extra salts, buffers from the sample because these molecules can greatly influence the data generated by LC-MS/MS.LC-MS/MS coupled with a mass spectrometer is one of the most important tools used to identify and quantify the immunoreactive cereal proteins (Alves et al., 2017). This coupling chromatographic separation and mass spectrometer detection techniques (LC-MS/MS) allows a large number of sampling in a short duration of time. Data generated by LC-MS/MS is large and it requires computational analytical effort to process the data in a systematic and comparative way in order to give a practical application (Victorio et al., 2018). Simultaneously, LC-MS/MS can be used as a confirmation method for various situations and certain levels. Therefore, LC-MS/MS has the future for the confirmation purpose rather than screening.

Though LC-MS/MS has many advantages over ELISA it also has some drawbacks such as enzyme choice for gluten hydrolysis can influence the results and converting this gluten from peptide is challenging. Simultaneously, LC-MS/MS is a very expensive tool and requires trained personnel along with expensive and specialized instrumentation (Alves et al., 2019).

### *7.7.8 Genomic Based Methods*

The genomic approach is a type of non-immunological method employed for the detection of offending food ingredients. ELISA, immunosensor, western blotting, LFD's all these immunological methods use antibody-antigen technique for the detection of food allergens. Whereas, genomic method basically relies on the amplification of DNA present in the food sample. This method serves as a marker to detect the food allergen. Polymerase chain reaction (PCR) is the only genomic method used till date to detect the gluten contents of food materials. PCR detects the specific DNA fragments in which primers are used to get the specificity that facilitate the amplification of the DNA. This DNA is species-specific and works as a marker to detect the food allergens of a particular food substance.

### 7.7.9 PCR

PCR is an alternative approach to ELISA and used for the quantification and identification of genetically modified organisms (GMO), pathogens, food allergens, etc. It is a common and recent DNA-based technology that became important when directives for GMO products got implemented. PCR generally targets the DNA of the allergens in the finished product (Popping, 2009). In PCR a very definite area of DNA is amplified by the primers. Primers are short stretches of DNA that bind selectively with the corresponding DNA which is to be amplified. These primers are protracted by using an enzyme called Taq Polymerase by the addition of nucleotides to make fragments of DNA. This technique continues several times to double the number of copies of the DNA fragment and it has been found that 30 numbers cycle can generate $10^9$ copies of DNA fragments. Then the result is separated using electrophoresis according to their size. The gel used is stained with a few fluorescent dyes so that when it intercalates with DNA it glows orange under ultra-violate (UV) light. Choice of primers is one of the important tasks to avoid false-positive results and it is generally performed to select the gliadin DNA (Besler, 2001).

At the time of using the PCR method cereal samples or the food products are selected and prepared and then the genomic DNA extraction is done. Then, the target gluten DNA sequence was selected for amplification. PCR is carried out and the products are purified. The sequencing reactions are performed by the primers using a PCR mixture containing the PCR buffers, NTP's (Nucleoside Triphosphates), and units of Taq polymerase. Thermal cycles such as denaturation of DNA followed by 30 cycles of denaturation then annealing followed by extension step are performed. Primers pair used for the amplification are selected based on their specificity. The amplified fragments are examined using electrophoresis in a 1.5% agarose gel for staining and carried out in a buffer. It was then visualized under UV light and a digital image was obtained (Martín-Fernăndez et al. 2015; Debnath et al., 2009).

Real-time PCR (RT-PCR) is the most commonly used PCR technique for the detection of allergenic foods. The RT-PCR does not include the use of gel in product detection. RT-PCR uses a specific probe with a reporter and quencher dye attached which joins the region flanked by oligonucleotides and primers and the enzyme reaction. The proximity of the quencher suppresses the fluorescence. During amplification, the 5′ exonuclease activity cleaves the hybridized probe and separates the dyes which results in an increase in fluorescence proportional to the amount of specific PCR product. The cycle number required until the fluorescence level dye is used to calculate the quantitative data (Poms et al., 2004). Real-time PCR makes the difference as they are based on melting curve analysis and specific for the cereals. In real-time PCR, gel electrophoresis is not used instead it is performed in a closed capillary for amplification which eliminates the risk of contamination. Real-time PCR based on melting curve analysis is less time-consuming with no requirement of gels and it can be used as a confirmatory tool in case of equivocal results since the melting behavior of PCR reflects the length and nucleotide sequence content.

But the problem with RT-PCR is it requires much expensive laboratory equipment than the conventional methods.

PCR is complementary to ELISA so both PCR and ELISA give reliable positive results and PCR allows identifying the positive results because there are no antibodies cross-reactivity (Immer & Haas-Lauterbach, 2010). While comparing applications of PCR with ELISA it has been found that PCR is more suitable for allergen quantification in complex food because PCR can detect the presence of DNA but not the protein, therefore the possibility of negative results are less as well as PCR have the advantages for rapid test development within 7–10 days if the DNA sequence is known, whereas, ELISA can take months for the same job. PCR relies on a simply defined DNA sequence while ELISA depends on animal antibody which require constant quality and stability. It has been found that PCR-ELISA combines the high specificity of the DNA-based method with a simple and economical ELISA method. The DNA fragment is amplified and bound to the surface of the microtiter well after denaturation of the DNA sequence probe is hybridized. The probe is a protein hapten labelled DNA fragment which is detected by the ELISA reaction and the quantitative data can be determined (Poms et al., 2004). Along with these advantages, PCR is cost-efficient, specificity is very high, and it can detect a very low amount of DNA by amplifying the analyte (gene-encoded allergenic protein). However, PCR has some disadvantages also such as PCR is a qualitative assay, PCR uses DNA and DNA does not dissolve in hydrophobic solution as protein does. PCR is more time-consuming and equipment-intensive than ELISA. Simultaneously, if DNA extracted from the food gets degraded then amplification becomes difficult followed by the failure of detection.

## 7.8 Novel Methods

In recent years various novel methods for gluten detection were reported such as magnetic beads, microarrays, multianalyte profiling, aptamers, etc. Magnetic beads use antigliadin polyclonal antibody-coated with magnetic beads which are used to detect gliadin present in food samples. These antigliadin polyclonal tagged liposomal nano vesicles encapsulated with fluorescent colour dye where IMBs (Immuno Magnetic Beads) gliadin forms sandwich complex followed by fluorescence detection. This immunoassay showed good recovery precision and sensitivity when the food samples (both processed and unprocessed food) were tested and the gliadin concentration or its fragments were determined and the results matched with R5 ELISA (Chu & Wen, 2013). These magnetic beads have the combined advantage of separation by magnetic beads and assays with sensitivity and robustness of detection. Reagents are used for rapid onsite and screen out the analysis of gliadin in foodstuff. Low cost, user-friendliness, mini size and compatibility, and mass fabrication possibilities make this technology attractive to be used for the rapid onsite analysis of food samples (Laube et al., 2011).

Microarrays are the multiplex lab chip-based device and the chip is incubated with capture agents such as fluorescent dye-labeled antibodies for gluten detection purposes. Microarrays are based on the immobilization technology of gliadin and chymotrypsin digested gliadin and fluorescent-labeled glutamine binding protein. Microarrays seemed promising but not yet tested (Cimaglia et al., 2014).

Multianalyte profiling is based on magnetic particles contain fluorescent dyes having different color codes. The proteins and peptides are conjugated to the surface and detection use specific antibody with phycoerythrin. This multiplex technique is used in 96 well microplates to detect 14 different food-based gluten. This device showed high sensitivity and specificity as well as labour requirement and time for analysis is less and also they are cheaper in cost. The cross-reactivity profiles of this device are clarified but still, further researches are required (Scherf & Poms, 2016).

Aptamers are a novel receptor for gluten detection which has now emerged profoundly. These are short DNA or RNA oligonucleotides that can detect a huge number of ligands with high specificity and selectivity. They are not much prone to thermal denaturation and they can be easily modified using any marker functionalities. They are generally raised against immune dominant peptides known as 33-mer. Aptamers targets complex foods that are present with a dissociation constant ($K_d$) in the nanomolecular range. The target molecules are highly specific because they can easily differentiate between closely arranged protein targets as well as aptamers are versatile and they can be performing against many targets (Osorio et al., 2019). The aptamers show neither negative false nor false-negative results. Its low value for reliable determination (3 mg/kg) would protect the sensitive celiac disease. This can be considered a step forward to limit the threshold of the labeled food in gluten-free countries up to 20 mg/kg value (López-López et al., 2017). Malvano et al. (2017) developed a new label-free impedimetric aptasensor for gluten detection based on immobilized aptamer (Gli 1). They modified it with PAMAM (Poly amidoamine-Dendrimer) which increased the sensitivity of the gluten detection limit to 5 ppm. An aptamer called gli 4 against 33 mer peptide was put in an electrochemical aptamer-based assay using magnetic practices followed by an examination of gliadin-containing samples were obtained (Amaya-González et al., 2015).

As discussed earlier aptamers assays are recent development methods for gluten quantification and it is still in a growing stage but it will undoubtedly advance. The detailed study and advancements may lead to modifying the aptamers with improved affinity and combination of aptamers with various nanomaterials such as carbon nanotubes with great variety and an easy to use tool for gluten detection will increase the acceptability of aptamers in near future (Miranda-Castro et al., 2016).

## 7.9 Variables and Challenges Influencing Gluten Detection

Scientists and researchers around the globe have developed various gluten detection techniques in the last decades. However, along with advantages, analytical methods developed to date have some limitations, as these developed methods cannot fully ensure affectability, selectivity, and exactness of a gluten reference. Hence, the advancements of analytical methods are required to develop for more sensitivity, reliability, accuracy, and cost-effectiveness (Miranda-Castro et al., 2016). Challenges to gluten detection are uniquely popularised due to the growing consumer demand for a gluten-free diet as gluten is directly linked to the celiac disease suffered by people all over the world which can be called gluten intolerance and for them, the only effective treatment is eliminating gluten completely from the diet (El Khoury et al., 2018). Consumer expectation has urged the food industry to develop analytical tools for gluten detection and also to produce a gluten-free diet. Therefore, there is a challenge for bakers, cereal researchers, and other industry people to produce gluten-free products without compromising the overall quality such as structure, texture, sensory qualities, and shelf life as that of gluten-containing products (Toufeili et al., 1994).

Gluten replacement is only possible if gluten can be easily detected and extracted from a food sample. Although epitopes (responsible for celiac disease) have been detected using the available analytical methods, still the structural complexity and polymorphism of gliadin proteins are required to be identified which is big a challenge for the researchers. Safety to gluten-free foods can only be identified by providing or developing more reliable methods of analysis. The current acceptance of R5 antibody ELISA and PCR are effective detection techniques but more sensitive and robust methods are required for the detection of gluten (Rosell et al., 2014). Available gluten detection methods work based on the quantification of alcohol-soluble prolamin fraction of gluten whereas the alcohol insoluble glutenin fraction is most of the times not targeted, whereas, both of the protein (prolamin/gliadin and glutelin) are containing immunogenic harmful peptides. Calculation of prolamin/gliadin factor is done by multiplying factor 2, based on the assumption that prolamin/gliadin and glutelin ratio is 1 and because of that and for some other errors most of the times the gluten content is overestimated which is a serious threat for celiac patients. Therefore, there is a need to develop a more sensitive, specific, robust technique for error-free suitable routine analysis purposes (Scherf & Poms, 2016).

## 7.10 Conclusion

The diseases and problems related to gluten have increased worldwide. Celiac disease is the prime pathological problem suffered by people due to gluten protein intake. The epitope concerned with celiac disease is recognized as the gliadin and glutenin and these fractions together form gluten. The complexity and polymorphism of these protein structures are difficult to identify. Currently, various countries around the world have passed resolutions and set a threshold limit for gluten-free and low gluten foods which are necessary to mention in the labeling of the food products. This will help the consumers to make safe food choices. Safety on gluten-free food is possible by providing reliable methods for gluten detection and quantification, whereas, it is still difficult to detect and extract gluten from food products mainly after processing.

However, a number of major improvements and advances in the analytical techniques of gluten since the last few decades have been observed. These improvements have certainly helped with the gluten-free labeling of the products which will ultimately help in increasing the food choices of celiac people. The current techniques employed and the legal compliance test gives the sensitivity required to follow codex limits up to 20 mg/kg. The globally permitted method is the ELISA method for the determination of gluten. Whereas competitive ELISA-based R5 antibody, monoclonal antibody method, G12method, and A1 method are also effective for the gluten determination. However, techniques such as ELISA along with PCR are facing problems like the detection is not possible for hydrolyzed food and it requires special extraction and tests. The proteomics method which is an alternative to PCR and ELISA has proved their specificity and sensitivity for food safety but they are expensive and required trained personnel for operation. Novel methods such as aptamers, microarrays are one of the processes of improvement for higher sensitivity and lower cross-reactivity and these techniques are growing fast as analytical tools. Novel methods with passing years are developing but still, they are suffering from the drawbacks of sensitivities and specificities as well as a further reduction in labour cost, infrastructure cost and recommendation from the Codex is also mandatory and should be taken care of.

Besides many drawbacks and advantages, all the above-mentioned methods and techniques for gluten detection require a large advancement to estimate gluten of the specific amount present. In the future, more advancements of these methods can be concerned for developing a collaborative method that will have all the advantages of the analytical methods employed till now. The use of the methods and its detection process mentioned in this chapter will give the readers a basic concept regarding the on-going methods employed and also help the forthcomings to develop a gluten-free diet and guide the world to have a better future by demolishing any type of diseases related to gluten and its risk factors.

## References

Alves, T.O., D'Almeida, C. T., & Ferreira, M. S. (2017). Determination of gluten peptides associated with celiac disease by mass spectrometry. *Celiac Disease and Non-Celiac Gluten Sensitivity, 43.*

Alves, T. O., D'Almeida, C. T., Scherf, K. A., & Ferreira, M. S. (2019). Modern approaches in the identification and quantification of immunogenic peptides in cereals by LC-MS/MS. *Frontiers in Plant Science, 10.*

Amaya-González, S., de los Santos Álvarez, N., Miranda-Ordieres, A. J., & Lobo-Castañón, M. J. (2015). Sensitive gluten determination in gluten-free foods by an electrochemical aptamer-based assay. *Analytical and Bioanalytical Chemistry*, 6021–6029.

Bean, S. R., & Lookhart, G. L. (2000). Electrophoresis of cereal storage proteins. *Journal of Chromatography*, 23–36.

Besler, M. (2001). Determination of allergens in foods. *Trends in Analytical Chemistry*, 662–672.

Bietz, J. A., & Wall, J. S. (1972). Wheat gluten subunits: Molecular weights determined by sodium dodecyl sulfate-polyacrylamide gel electrophoresis.

Camafeita, E., & Méndez, E. (1998). Screening of gluten avenins in foods by matrix-assisted laser desorption/ionization time-of-flight mass spectrometry. *Journal of Mass Spectrometry*, 1023–1028.

Cao, W., Watson, D., Bakke, M., Panda, R., Bedford, B., Kande, P. S., Jackson, L. S., & Garber, E. A. (2017). Detection of gluten during the fermentation process to produce soy sauce. *Journal of Food Protection, 80*(5), 799–808.

Catassi, C., & Yachha, S. K. (2016). *Science of gluten-free foods and beverages* (E. Arendt, & F. Dal Bello, Eds., pp. 1–11). Academic.

Catassi, C., Fabiani, E., Rätsch, I. M., Coppa, G. V., Giorgi, P. L., Pierdomenico, R., Alessandrini, S., Iwanejko, G., Domenici, R., Mei, E., & Miano, A. (1996). The coeliac iceberg in Italy. A multicentreantigliadin antibodies screening for coeliac disease in school-age subjects. *Actapaediatrica*, 29–35.

Cho, C. Y., Nowatzke, W., Oliver, K., & Garber, E. A. (2015). Multiplex detection of food allergens and gluten. *Analytical and Bioanalytical Chemistry, 407*(14), 4195–4206.

Chu, P. T., & Wen, H. W. (2013). Sensitive detection and quantification of gliadin contamination in gluten-free food with immunomagnetic beads based liposomal fluorescence immunoassay. *Analytica Chimica Acta*, 246–253.

Chu, P. T., Lin, C. S., Chen, W. J., Chen, C. F., & Wen, H. W. (2012). Detection of gliadin in foods using a quartz crystal microbalance biosensor that incorporates gold nanoparticle. *Journal of Agriculture and Food Chemistry*, 6483–6492.

Cimaglia, F., Potente, G., Chiesa, M., Mita, G., & Bleve, G. (2014). Study of a new gliadin capture agent and development of a protein microarray as a new approach for gliadin detection. *Journal of Proteomics and Bioinformatics*, 248–255.

Debnath, J., Martin, A., & Gowda, L. R. (2009). A polymerase chain reaction directed to detect wheat glutenin: Implications for gluten-free labelling. *Food Research International*, 782–787.

Denham, J. M., & Hill, I. D. (2013). Celiac disease and autoimmunity: Review and controversies. *Current Allergy and Asthma Reports*, 347–353.

Diaz-Amigo, C. (2010). Antibody-based detection methods: From theory to practice. In *Molecular biological and immunological techniques and applications for food chemists* (pp. 223–245). Wiley.

Dostalek, P., Gabrovska, D., Rysova, J., Mena, M. C., Hernando, A., Méndez, E., Chmelik, J., & Šalplachta, J. (2009). Determination of gluten in glucose syrups. *Journal of Food Composition and Analysis*, 762–765.

DuPont, F. M., Chan, R., Lopez, R., & Vensel, W. H. (2005). Sequential extraction and quantitative recovery of gliadins, glutenins, and other proteins from small samples of wheat flour. *Journal of Agricultural and Food Chemistry*, 1575–1584.

El Khoury, D., Balfour-Ducharme, S., & Joye, I. J. (2018). A review on the gluten-free diet: Technological and nutritional challenges. *Nutrients, 1410.*

Gomaa, A., & Boye, J. I. (2013). Impact of thermal processing time and cookie size on the detection of casein, egg, gluten and soy allergens in food. *Food Research International*, 483–489.

Haas-Lauterbach, S., Immer, U., Richter, M., & Koehler, P. (2012). Gluten fragment detection with a competitive ELISA. *Journal of AOAC international*, 377–381.

Haraszi, R., Chassaigne, H., Maquet, A., & Ulberth, F. (2011). Analytical methods for detection of gluten in food—Method developments in support of food labeling legislation. *Journal of AOAC International*, 1006–1025.

Harlow, E., & Lane, D. (1988). *Antibodies: A laboratory manual*. Cold Spring Harbor Laboratory Press.

Harlow, E., & Lane, D. (1998). *Using antibodies: A laboratory manual*. Cold Spring Harbor Laboratory Press.

Hattersley, S., Ward, R., Baka, A., & Crevel, R. W. (2014). Advances in the risk management of unintended presence of allergenic foods in manufactured food products – An overview. *Food and Chemical Toxicology*, 255–261.

Hernando, A., Mujico, J. R., Mena, M. C., Lombardía, M., & Mendez, E. (2008). Measurement of wheat gluten and barley hordeins in contaminated oats from Europe, the United States and Canada by Sandwich R5 ELISA. *European Journal of Gastroenterology & Hepatology*, 545–554.

Howard, G. C., & Bethell, D. R. (Eds.). (2000). *Basic methods in antibody production and characterization*. CRC Press.

Immer, U., & Haas-Lauterbach, S. (2010). Gluten detection (Chapter 19). In B. Popping, C. Diaz-Amigo, & K. Hoenicke (Eds.), *Molecular biological and immunological techniques and applications for food chemists* (pp. 359–375). Wiley.

Kanerva, P. M., & Sontag-Strohm, T. S., (2006). Problems in detecting prolamins contaminants in oat-based foods by commercial ELISA kits. In *9th international gluten workshop,* September (p. 33).

Kanerva, P., Brinck, O., Sontag-Strohm, T., Salovaara, H., & Loponen, J. (2011). Deamidation of gluten proteins and peptides decreases the antibody affinity in gluten analysis assays. *Journal of Cereal Science*, 335–339.

Khan, K., & Huckle, L. (1992). Use of multistacking gels in sodium dodecyl sulfate-polyacrylamide gel electrophoresis to reveal polydispersity, aggregation, and disaggregation of the glutenin protein fraction. *Cereal Chemistry*, 686–688.

Kolster, P., Krechting, C. F., & Van Gelder, W. M. J. (1992). Quantification of individual high molecular weight subunits of wheat glutenin using SDS—PAGE and scanning densitometry. *Journal of Cereal Science*, 49–61.

Laube, T., Kergaravat, S. V., Fabiano, S. N., Hernández, S. R., Alegret, S., & Pividori, M. I. (2011). Magneto immunosensor for gliadin detection in gluten-free foodstuff: Towards food safety for celiac patients. *Biosensors and Bioelectronics*, 46–52.

Lee, H. J., Anderson, Z., & Ryu, D. (2014). Gluten contamination in foods labeled as "gluten free" in the United States. *Journal of Food Protection*, 1830–1833.

Leszczynska, J., Łącka, A., Szemraj, J., Lukamowicz, J., & Zegota, H. (2003). The influence of gamma irradiation on the immunoreactivity of gliadin and wheat flour. *European Food Research and Technology*, 143–114.

Liao, Y. S., Kuo, J. H., Chen, B. L., Tsuei, H. W., Lin, C. Y., Lin, H. Y., & Cheng, H. F. (2017). Development and validation of the detection method for wheat and barley glutens using mass spectrometry in processed foods. *Food Analytical Methods, 10*(8), 2839–2847.

Lock, S. (2014). Gluten detection and speciation by liquid chromatography mass spectrometry (LC-MS/MS). *Food*, 13–29.

Lock, S., Lane, C., Jackson, P., & Serna, A. (2010). *The detection of allergens in bread and pasta by liquid chromatography tandem mass spectrometry*. Application Note SCIEX.

Lookhart, G. L., & Wrigley, C. W. (1995). *Identification of food grain varieties* (C.W. Wrigley, Ed., pp. 55–71). AACC International.

López-López, L., Miranda-Castro, R., de Los Santos Alvarez, N., Miranda-Ordieres, A. J., & Lobo-Castañón, M. J. (2017). Disposable electrochemical aptasensor for gluten determination in food. *Sensors and Actuators B: Chemical*, 522–527.

Malvano, F., Albanese, D., Pilloton, R., & Di Matteo, M. (2017). A new label-free impedimetricaptasensor for gluten detection. *Food Control*, 200–206.

Marín-Barroso, E., Messina, G. A., Bertolino, F. A., Raba, J., & Pereira, S. V. (2019). Electrochemical immunosensor modified with carbon nanofibers coupled to a paper platform for the determination of gliadins in food samples. *Analytical Methods*, 2170–2178.

Martín-Fernández, B., Costa, J., Oliveira, M. B. P., López-Ruiz, B., & Mafra, I. (2015). Screening new gene markers for gluten detection in foods. *Food Control, 56*, 57–63.

Martín-Yerga, D., & Costa-García, A. (2014). Electrochemical immunosensors for celiac disease detection. *International Journal*, 139–141.

Melini, F., & Melini, V. (2018). Immunological methods in gluten risk analysis: A snapshot. *Safety, 56.*

Mena, M. C., Lombardía, M., Hernando, A., Méndez, E., & Albar, J. P. (2012). Comprehensive analysis of gluten in processed foods using a new extraction method and a competitive ELISA based on the R5 antibody. *Talanta, 91*, 33–40.

Miranda-Castro, R., De-los-Santos-Álvarez, N., Miranda-Ordieres, A. J., & Lobo-Castañón, M. J. (2016). Harnessing aptamers to overcome challenges in gluten detection. *Biosensors, 16.*
Mothes, T., & Stern, M. (2003). How gluten-free is gluten-free, and what does this mean to coeliac patients? *European Journal of Gastroenterology & Hepatology, 15*(5), 461–463.
Nassef, H. M., Bermudo Redondo, M. C., Ciclitira, P. J., Ellis, H. J., Fragoso, A., & O'Sullivan, C. K. (2008). Electrochemical immunosensor for detection of celiac disease toxic gliadin in foodstuff. *Analytical Chemistry*, 9265–9271.
Neves, M. M., González-García, M. B., Nouws, H. P., & Costa-García, A. (2012). Celiac disease detection using a transglutaminase electrochemical immunosensor fabricated on nanohybrid screen-printed carbon electrodes. *Biosensors and Bioelectronics*, 95–100.
Newberry, C., McKnight, L., Sarav, M., & Blakely, O. P. (2017). Going gluten free: The history and nutritional implications of today's Most popular diet. *Current Gastroenterology Reports, 54.*
Niewinski, M. M. (2008). Advances in celiac disease and gluten free diet. *Journal of American Dietetic Association*, 661–672.
Osorio, C. E., Mejías, J. H., & Rustgi, S. (2019). Gluten detection methods and their critical role in assuring safe diets for celiac patients. *Nutrients, 2920.*
Panda, R., & Garber, E. A. (2019). *Detection and quantitation of gluten in fermented-hydrolyzed foods: Challenges, progress and potential path forward* (p. 97). Frontiers in Nutrition.
Panda, R., Boyer, M., & Garber, E. A. (2017). A multiplex competitive ELISA for the detection and characterization of gluten in fermented-hydrolyzed foods. *Analytical and Bioanalytical Chemistry*, 6959–6973.
Park, S. H., Bean, S. R., Chung, O. K., & Seib, P. A. (2006). Levels of protein and protein composition in hard winter wheat flours and the relationship to breadmaking. *Cereal Chemistry*, 418–423.
Payne, P. I., Corfield, K. G., & Blackman, J. A. (1979). Identification of a high-molecular-weight subunit of glutenin whose presence correlates with bread-making quality in wheats of related pedigree. *Theoretical and Applied Genetics*, 153–159.
Pedersen, R. O., Nowatzke, W. L., Cho, C. Y., Oliver, K. G., & Garber, E. A. (2018). Cross-reactivity by botanicals used in dietary supplements and spices using the multiplex xMAP food allergen detection assay (xMAP FADA). *Analytical and Bioanalytical Chemistry*, 5791–5806.
Poms, R. E., Anklam, E., & Kuhn, M. (2004). Polymerase chain reaction techniques for food allergen detection. *Journal of AOAC International*, 1391–1397.
Popping, B. (2009). Challenges in detecting food allergens—Analytical methods in the legal context. In *The Science of Gluten-Free Foods and Beverages*. AACC International Press.
Righetti, P. G., & Bosisio, A. B. (1981). Applications of isoelectric focusing to the analysis of plant and food proteins. *Electrophoresis*, 65–75.
Rosell, C. M., Barro, F., Sousa, C., & Mena, M. C. (2014). Cereals for developing gluten-free products and analytical tools for gluten detection. *Journal of Cereal Science*, 354–364.
Saghatelian, A., & Cravatt, B. F. (2005). Global strategies to integrate the proteome and metabolome. *Current Opinion in Chemical Biology*, 62–68.
Sandberg, M., Lundberg, L., Ferm, M., & Yman, I. M. (2003). Real time PCR for the detection and discrimination of cereal contamination in gluten free foods. *European Food Research and Technology*, 344–349.
Scherf, K. A., & Poms, R. E. (2016). Recent developments in analytical methods for tracing gluten. *Journal of Cereal Science*, 112–122.
Sharma, G. M., & Rallabhandi, P. (2015). The effects of processing on gluten from wheat, rye, and barley, and its detection in foods. In *Processing and impact on active components in food*. Academic.
Sharma, G. M., Pereira, M., & Williams, K. M. (2014). Gluten detection in foods available in the United States - A market survey. *Food Chemistry*, 120–126.
Sharma, G. M., Pereira, M., & Williams, K. M. (2015). Gluten detection in foods available in the United States–A market survey. *Food Chemistry*, 120–126.
Shefcheck, K. J., & Musser, S. M. (2004). Confirmation of the allergenic peanut protein, Ara h 1, in a model food matrix using liquid chromatography/tandem mass spectrometry (LC/MS/MS). *Journal of Agricultural and Food Chemistry*, 2785–2790.

Shewry, P. R. (2007). Improving the protein content and composition of cereal grain. *Journal of Cereal Science*, 239–250.

Shewry, P. R. (2009). Wheat. *Journal of Experimental Botany*, 1537–1553.

Shewry, P. R., & Halford, N. G. (2002). Cereal seed storage proteins: Structures, properties and role in grain utilization. *Journal of Experimental Botany*, 947–958.

Skerritt, J. H., & Hill, A. S. (1990). Monoclonal antibody sandwich enzyme immunoassays for determination of gluten in foods. *Journal of Agriculture Food Chemistry*, 1771–1778.

Skerritt, J. H., Diment, J. A., & Wrigley, C. W. (1985). A sensitive monoclonal – Antibody- based test for gluten detection: Choice of primary and secondary bodies. *Journal of Science and Food Agriculture*, 995–1003.

Slot, I. D. B., Fels-Klerx, H. J. V. D., Bremer, M. G. E. G., & Hamer, R. J. (2016). Immunochemical detection methods for gluten in food products: Where do we go from here. *Critical Reviews in Food Science and Nutrition*, 2455–2466.

Thompson, T., & Méndez, E. (2008). Commercial assays to assess gluten content of gluten-free foods: Why they are not created equal. *Journal of the American Dietetic Association*, 1682–1687.

Toufeili, I. M. A. D., Dagher, S. H. A. W. K. Y., Shadarevian, S. O. S. S. Y., Noureddine, A. B. I. R., Sarakbi, M., & Farran, M. T. (1994). Formulation of gluten-free pocket-type flat breads: Optimization of methylcellulose, gum Arabic, and egg albumen levels by response surface methodology. *Cereal Chemistry*, 594–600.

Victorio, V. C. M., Souza, G. H., Santos, M. C. B., Vega, A. R., Cameron, L. C., & Ferreira, M. S. L. (2018). Differential expression of albumins and globulins of wheat flours of different technological qualities revealed by nanoUPLC-UDMSE. *Food Chemistry*, 1027–1036.

Weng, X., Gaur, G., & Neethirajan, S. (2016). Rapid detection of food allergens by microfluidics ELISA-based optical sensor. *Biosensors, 6*(2), 24.

Werner, W. E., Wiktorowicz, J. E., & Kasarda, D. D. (1994). Wheat varietal identification by capillary electrophoresis of gliadins and high molecular weight glutenin subunits. *Cereal Chemistry*, 397–402.

Wieser, H. (2003). The use of redox agents. In S. P. Cauvain (Ed.), *Bread making-improving quality* (pp. 424–446). Woodhead Publishing Ltd.

Wieser, H. (2008). Detection of gluten. Chapter 3. In E. K. Arendt & F. Dal Bello (Eds.), *Gluten-free cereal products and beverages* (pp. 47–80). Academic Press.

Windemann, H., Fritschy, F., & Baumgartner, E. (1982). Enzyme-linked immunosorbent assay for wheat α-gliadin and whole gliadin. *Biochimicaet Biophysica Acta (BBA)-Protein Structure and Molecular Enzymology*, 110–121.

Woychik, J. H., Boundy, J. A., & Dimler, R. J. (1961). Starch gel electrophoresis of wheat gluten proteins with concentrated urea. *Archives of Biochemistry and Biophysics*, 477–482.

Wrigley, C. W. (1996). Giant proteins with flour power. *Nature*, 738–739.

Xhakollari, V., Canavari, M., & Osman, M. (2018). Factors affecting consumer's adherence to gluten free diet, a systematic review. *Trends in Food Science and Technology*, 23–33.

Yman, I. M., Eriksson, A., Everitt, G., Yman, L., & Karlsson, T. (1994). Analysis of food proteins for verification of contamination or mislabelling. *Food and Agricultural Immunology*, 167–172.

Zhang, J., Portela, S. B., Horrell, J. B., Leung, A., Weitmann, D. R., Artiuch, J. B., Wilson, S. M., Cipriani, M., Slakey, L. K., Burt, A. M., & Lourenco, F. J. D. (2019). An integrated, accurate, rapid, and economical handheld consumer gluten detector. *Food Chemistry*, 446–456.

Žilić, S. (2013). Wheat gluten: Composition and health effects. *Gluten, 71-86.*

Zuidmeer, L., Goldhahn, K., Rona, R. J., Gislason, D., Madsen, C., Summers, C., Sodergren, E., Dahlstrom, J., Lindner, T., Sigurdardottir, S. T., & McBride, D. (2008). The prevalence of plant food allergies: A systematic review. *Journal of Allergy and Clinical Immunolog*, 1210–1218.

# Chapter 8
# Novel Approaches in Gluten-Free Bread Making: Case Study

**E. J. Rifna, Madhuresh Dwivedi, and Rewa Kulshrestha**

**Abstract** Celiac disease is the most commonly reported human chronic gastrointestinal disease. The unique effectual therapy for victims with celiac disease is to pursue a diet free of gluten strictly. Currently, the rising occurrence of celiac diseases encourages global attentiveness for diverse favored gluten-free products. Therefore, the increasing requirement for high-quality gluten-free bread from natural compounds is increasing the want for novel approaches in gluten-free breadmaking. Nevertheless, baking devoid of gluten, the chief component for bread texture, quality, and structure, is a great confront for every confectioner and cereal researchers. Various methods have been used to comprehend and develop a gluten-free bread system by monitoring various starch properties, flour sources, additives, and the use of technology or synergistic effect of these elements. Few works intended to evaluate or progress gluten-free bread technical or dietary attributes, whereas others aimed at manifold objectives. Some studies applied food science elements to develop the sensory property of gluten-free bread, mutually with nutritional aspects. Henceforth, the important focus of this book chapter is to confer the new approaches for gluten-free bread improvements in the past few years, including sourdough, the role of hydrocolloids, innovative techniques, and nutritional enhancement.

**Keywords** Celiac disease · Bread · Nutritional quality · Hydrocolloids · Prebiotics

E. J. Rifna · M. Dwivedi (✉)
Department of Food Process Engineering, NIT Rourkela, Rourkela, Odisha, India
e-mail: dwivedim@nitrkl.ac.in

R. Kulshrestha
Department of Food Processing and Technology, Atal Bihari Vajpayee Vishwavidyalaya, Bilaspur, Chhattisgarh, India


N. Singh Deora et al. (eds.), *Challenges and Potential Solutions in Gluten Free Product Development*, Food Engineering Series,
https://doi.org/10.1007/978-3-030-88697-4_8

## 8.1 Introduction

Celiac disease is a dreadful autoimmune disease spotted by enduring intolerance to compounds says gliadin, hordein, secalin, and avidins in wheat, barley, rye, and oats respectively, owing to genetic characteristics (Mohammadi et al., 2014). Celiac disease leads to immunologically related inflammatory damage of the mucosa layer in the small intestine that results in malabsorption of essential food ingredients and gastrointestinal problems (Kagnoff, 2005). Presently, works have revealed that celiac disease attacks mostly half percent of the global populace. The sole present remedy is a lifetime complete elimination of gluten and other associated prolamines from a daily diet. In before years, European Union formulated a guideline explaining that gluten-free foods that are composed of naturally gluten-free compounds should only hold gluten in an amount less than 20 ppm (Demirkesen et al., 2010; Deora et al., 2014). Gluten is the chief structure forming compound in wheat bread that gives out the dough its distinctive rheological properties and baking quality.

Contrast to bread dough with gluten compound, in gluten-free bread dough the stereoscopic structured protein-starch complex is absent, and they are majorly prepared from refined flour or unadulterated starches (e.g. corn starch, rice flour). To permit the starch-rich compounds to completely gelatinize through baking and to enhance the viscosity and thus improve the gas-holding property, significantly greater water contents are essential in gluten-free formulations. This considerably alters the dough consistency towards a batter which negatively modifies the production parameters and the final bread quality. Henceforth owing to its inimitable functionality, the substitution of gluten stands to be challenging. Furthermore, the distinctive attributes of wheat gluten make it cumbersome to discover raw ingredients, or additives, which could completely substitute it and currently, many gluten products accessible in the market have reduced dietary properties, penurious taste, and poor quality (Mustalahti et al., 2010; Tripathi et al., 2017). Various approaches have been used to assist in processing and improve gluten-free bread attributes. The majority of been these was based on applying multifaceted formulations comprising of a grouping of various additives and ingredients, so as to imitate the gluten structure. Regardless through various works, no single baking additive was significantly found capable to replicate the gluten network to its full potential yet (Niewinski, 2008; Mishra et al., 2020).

During the last few years, the addition of new substitutive ingredients involving starches, fibers, proteins, emulsifiers, enzymes, and gluten-free flours (López-Tenorio et al., 2015; Tsatsaragkou et al., 2014; Wronkowska et al., 2013; Ziobro et al., 2016), was found promising in improving dough rheological characteristics, aiding processing parameters and improving the nutritional profile. More lately, novel approaches say sourdough fermentation, physical treatments, prebiotic gluten-free bread, and partial baking technology (Basso et al., 2015; Stefańska et al., 2016; Jerome et al., 2019) have been demonstrated to be a favorable alternative substitute to develop gluten-free bread of significant quality. This chapter focuses on the novel approaches that have been undertaken towards the development of high-quality gluten-free bread.

## 8.2 Sourdough in Gluten-Free Bread Baking

Sourdough is a blend of flour and water agitated with yeast and lactic acid bacteria (LAB), that decide its properties with respect to aroma and production of acids. The application of sourdough has an extended tradition in the baking of rye and wheat bread (Gänzle et al., 2008). In particular, fermentation developed by lactic acid bacteria is a precondition for rye bread preparation, as it enhances the dissolution of rye pentosans, which decide the rye bread texture, structure, and this also ceases the activity of amylase (Alvarez-Jubete et al., 2009). When added in measured proportions, sourdough improves texture, volume, and health aspects by enhancing the shelf life of bread by protecting it from mold spoilage. These affirmative effects are related to the metabolic behavior of sourdough-resident microbes, like exopolysaccharides (EPS) production and release of antimicrobial compounds (De Vuyst & Vancanneyt, 2007).

The use of sourdough is an old method that has been applied for a long time and is achieving attention over again. Few judges it therefore to be a "novel" methodology (Moroni et al., 2009). In recent days consumers who insist on clean labels have rerouted the center of study in identifying alternative tools that permit the development of elevated quality of gluten-free bread without the use of additives. Decreasing food additives quantity could decrease extreme ingredient price and eliminate the occurrence of few allergens in the final baked product. Current investigations revealed that when applied in the correct ratio, sourdough can be used to tackle the majority of the issues related to the baking of low-quality gluten-free bread, whilst being and echo friendly and cost effectual (Cappa et al., 2016). The positive property is related to the occurrence of few by-products produced from LAB, say antimicrobial molecules, EPS, volatiles, and lactic acid which are formed at the process of fermentation. Dough acidification could also trigger few endogenous flour chemical molecules such as enzymes which can later makes bread crumb softer. The ratio among acids is necessary, as it determines the texture and structure of bread (Arendt et al., 2007). The acetic acid and lactic acid also increase the shelf life of bread, as it prolongs staling by disturbing retrogradation of starch.

With respect to the microbial constituents, works emphasized the significance of choosing suitable starter strains for the preparation of gluten-free sourdoughs, as every microorganism cannot adjust evenly to the identical raw compound. Microbial growth can also get affected by the accessibility of carbohydrates, lipids, nitrogen content, and percentage of free fatty acids, in addition to the buffer capacity and enzymatic activity in the substrate (Moroni et al., 2010). Nevertheless, the property and superiority of the raw ingredient are not the sole properties that assess the sourdough microbiota. Parameters such as dough yield, fermentation temperature, time and autochthonous culture also influence the end constituents of the sourdough

(Arendt et al., 2007; Rifna et al., 2019). *L. Plantarum* and *Lactobacillus fermentum*, strain commonly separated from gluten-free sourdoughs from, rice, amaranth, and teff (Moroni et al., 2009). Between LAB used, *Lactobacillus plantarum* is the main reported in gluten-free sourdoughs prepared from quinoa, rice, and amaranth (Moroni et al., 2010). The authors identified that the above strain released organic acids (lactic acid) which were antifungal (Moore et al., 2008), and enhanced firmness of crumb and staling level of gluten-free bread prepared with the refined flour (Moore et al., 2007). The use of sourdough was identified to enhance the textural properties as studied by (Houben et al., 2010). Likewise, (Jekle et al., 2010) reported that the incorporation of amaranth sourdough appreciably affected the rheological characteristics of amaranth batters and these effects were reliant on the quantity of sourdough added used in fermentation process.

Another work explained that sorghum sourdough fermented with *most lactobacillus strain* improved final bread nutritional value by depressing the polyphenolic activity (Svensson et al., 2010). As aforementioned, the microbiota in sourdough will assess the dough attributes regards to the aroma, acidification, and leavening. Few LAB strains release EPS that increase the elastic characteristic of batter, but also enhance the structure and bread shelf-life. The major general EPS employed are fructan, levan, and dextran. Sourdough is an important assuring technique to be used at the time of gluten-free bread preparation as it enhances concurrently nutritional and sensorial final bread attributes. Important is the assortment of suitable starter strains that should be cautiously selected for every particular raw ingredient. Notably, four important parameters are dependable for the supremacy of lactic acid bacillus strains: the compliance of sugar metabolism, pH, fermentation temperature, and the release of the antimicrobial molecule. These parameters will donate to the perseverance of the superior cultures and could make certain a reproducible and restricted ratio of sourdough microbiota and guarantee an even final bread quality.

## 8.3 Aeration Strategies

To evaluate different aeration methods the level of gas entrapment has to be defined which is often challenging. The different methods and formulas which have been employed to describe the amount of air in the dough will be discussed below about their applicability for gluten-free dough. Various methods have been demonstrated to identify the gas entrapment of bread. The chief objective of these techniques is to know how the gas is dispersed among the continuous phase. It has been reported that two different dough samples holding equal gas content can possess dissimilar crumb textures, owing to the dissimilarity regarding the size distribution and their primary bubble sizes (Chin & Campbell, 2005). The dullness and volatility of dough create complications to identify the arrangement and the characteristics of formed gas bubbles. As the gluten-free dough is regularly more flowing and less elastic when compared to dough out of wheat, disproportionation and buoyancy could play a greater part in its foam constancy. Henceforth, appropriate technologies for assessing gas bubbles all through processing in the gluten-free dough are mandatory. However various aeration methods, applied for gluten-free bread are described below.

## 8.4 Biological Aeration

To attain a fine crumb grain structure with a large volume, the bubble amount and size added at the process of mixing have to enhance uniformly in the succeeding processing steps. For this reason, biological aeration using microorganisms, say yeast, is suitable, as it constantly releases carbon dioxide until the neighboring circumstances are positive. The amount of carbon dioxide developed relies on the viability of the chosen microorganism, ionic strength, pH, temperature, substrate, humidity, and the availability of the nutrients. The ensuing development of bubbles at the process of proofing and baking can be modeled (Chiotellis & Campbell, 2003). To produce an accurate computation of the dispersion of carbon dioxide into air bubbles it seems demanding to monitor its quantity in the liquid dough state. From the outcomes, it was estimated that the end bubble dimension relies on the original size at the commencement of the fermentation process (Chiotellis & Campbell, 2003). This explains the significance of the primary mixing methods for creating minute nuclei with a fine size distribution. An appropriate model valid to gluten-free dough might be beneficial for understanding the correct option of baking and fermentation process.

## 8.5 Chemical Aeration

Another strategy for better aeration of gluten-free bread is the application of chemical raising compounds as a substitute or in couple to baker's yeast. Alike to biological leavening, the aeration using chemical agents relies on the amount of carbon dioxide generated and the capability of the gluten-free dough to hold the gas. In the case of wheat bread, the use of chemical raising agents is very limited, they are used for numerous gluten-free foodstuffs in addition to added microbes. Segura and Rosell (2011) evaluated gluten-free bread procured from the market which integrated chemical raising agents. Confirmative, (Sinelli et al., 2008) demonstrated that added chemical agents are frequent in wheat and gluten-free bread composition. It is noteworthy to cautiously select the kind, combination baking powder with the precise characteristics and necessities of the product in mind. Henceforth, it is startling that to date no valid research works concerning the correct alternative of chemical leavening substitutes are developed.

## 8.6 Physical Aeration

Chemical and biological aeration are generally followed by an agitating process that pretends a particular role for the ensuing bubble distribution. With respect to the quantity and arrangement of original gas nuclei, the rate of the successive bubble development and constancy at the period of baking gets affected. This could be the major significant objective for an alteration of conventional processing steps to the necessities of gluten-free dough. In early 1962, a substitute for conventional kneading was introduced, by coupling elevated speed mixing using the addition of vacuum (Cauvain & Young, 2006). As this method permits for an indirect inflection of bubble sizes by altering the head pressure, there could be possible for an amendment to the necessities of gluten-free dough. Massey et al. (2001) identified that an augment of the vacuum at the process of mixing boosted aeration and decreased the bubble dimensions. In this work though, the enhanced gas volume was owing to the development of the bubbles after pressure release and not owing to improved gas retentiveness. Usually, pressure development creates fine bubbles that enlarge as fast as the pressure is set free which could retain ingredients and process time. Cheng (1992) demonstrated a patented technique to join a mixer and an ultrasonic bath for an enhancement of cake batter aeration using acoustic cavitation. The aeration with aid of ultrasound was studied as a superior technique for aeration of batter at the laboratory level, whereas its incorporation at the industrial level could strongly increase operating expenses (Chin et al., 2015). In general, the adaption of the aeration strategies discussed for gluten-free bread seems to be a potential tool. However, future experiments that evaluate the effect of mixing and the gas volume fraction non-destructively on gluten-free dough need to be developed.

The most used tool for evaluating aeration strategies in food molecules is the application of the image analysis technique. With respect to the cake batter, a charge-coupled device camera was coupled with microscopy for monitoring gas bubbles (Hicsasmaz et al., 2003). Through another work, researchers examined physically divide samples of gluten-free frozen dough under cryo-scanning electron microscopy for better resolution (Trinh et al., 2013). Authors found that freezing affected the density of dough and also cell arrangement was reported to be distorted (Campbell & Mougeot, 1999). Later, (Trinh et al., 2013) demonstrated that starch granules could have been misplaced at the period of fracturing operation and their remaining could be misguided for air bubbles. A synopsis involving works applying microscopy has been provided by (Campbell & Martin, 2012), who explained significant variations regards to the bubble sizes (35–112 μm), gas volume fraction (3.5–10%), and a robust correlation on the experiential slice thickness of gluten-free products. Confocal scanning microscopy aids a three-dimensional revelation of the grain structure of bread dough after streaking particular ingredients (Jekle & Becker, 2011). For the gas bubble assessment, the ingredients of the neighboring medium have to be examined. Inadequately colored samples result in producing smaller bubbles, which can forge the outcome (Richardson et al., 2002). Similarly, interactions among dyes, gluten-free dough, and normal dough molecules could have an impact on the observable texture. Gas bubbles of bigger size (500–2000 μm) are complicated to observe through computer scanning microscopy as they hold the majority or the entirety of the visible region.

## 8.7 Nutritional Enhancement

Dietary fibers have been extensively considered for their functional properties in gluten-free bread composition, with regards to their water fastening property, fat mimetic properties, gel-forming capacity, and textural effects (Wang et al., 2017). Researches were performed to study the impacts of insoluble fibers on the sensorial property of gluten-free bread (Utrilla-Coello et al., 2013). Dough consistency and pasting characteristics of starch were also found to be affected fractionally by incorporation of fiber (oat bran) (Aprodu & Banu, 2015), due to their significant water-binding ability to present dough rheology and gelatinization of starch for gluten-free bread making (Demirkesen et al., 2010).

The addition of starch and soluble fiber compounds may lower the glycemic response of gluten-free bread that is significantly advantageous for folks with concurrent celiac disease and diabetes. Preparing composition with functional fibers for example psyllium and β-glucan have been researched widely as a remedy to aid regulation of gut and reduce serum low-density lipoprotein cholesterol values (Gunness & Gidley, 2010).

Prebiotics says oligofructose, resistant starch, and inulin are the most commonly studied functional dietary fibers for gluten-free bread preparation. As per (Capriles & Arêas, 2013), gluten-free bread with a high percentage (4–12%) of inulin-type fructans (ITFs) exhibited specific volume less than 10%, whilst noticing a turn down higher than 10%. The authors recommended that ITFs could develop a gel arrangement and hold carbon dioxide similar to few hydrocolloids. Various degrees of polymerization of inulin also produce a significant impact on the bread quality. Usually, a reduced degree of polymerization of inulin has robust effects compared to superior ones (Ziobro et al., 2013). Resistant starch develops numerous functional properties and could not only lower the food energy but also improves digestive properties and final bread properties (Witczak et al., 2016). Furthermore, resistant starch does not affect the bread crumb firmness whereas enhances its rheological properties particularly, porosity and elasticity (Tsatsaragkou et al., 2014).

## 8.8 Changing Flour Functionality Through Physical Treatments

Gluten-free flour could be physically altered using various particle size categorization and milling techniques. On one side these physical treatments are used to stabilize gluten-free flour and enhance shelf life whereas on the other side novel functionalities are developed. Henceforth, the flour developed after these physical treatments varies in its properties, say thickening capacity, water binding ability, pasting properties, emulsifying characteristics, and chemical activity towards proteins, enzymes, and others.

### *8.8.1 Particle Size Classification*

In this approach, (Kadan et al., 2008) and (Araki et al., 2009) demonstrated that the milling process affected the broken starch and particles of refined flour and henceforth the bread volume was greatly affected. Through the above work, the authors observed a significant negative trend among broken starch and a specific volume of the final bread. On the other hand, the authors must incorporate wheat gluten into the bread recipe, so that outcomes cannot be entirely extrapolated. Whereas, (de la Hera et al., 2013b) observed that as there was a decrease in particle size of refined flour the specific volume also decreased for gluten-free flour. This impact was credited to the characteristics of dough at the fermentation process, as dough prepared with flours were barely capable of preserving gas released, which could be owing to the structural variations demonstrated among various doughs. Nevertheless, variations in the broken starch between the flours categorized through sieving were diminutive and, contrary to the outcome expected, the best portion was that which exhibited a reduced percentage of damaged starch (de la Hera et al., 2013a). Henceforth, the starch damaged itself could not elucidate these variations on gluten-free bread volume.

By working on semi-dry milled refined flours accessed through air classification, (Park et al., 2014) demonstrated that the superior portions produced bread of reduced volume, though a reduced starch percentage (<5%) was existing. In general, it has been demonstrated with rice flour that, finer flours baked bread of reduced specific volume (de la Hera et al., 2013c). Also, regards to oat bread, oat flours possessing coarse particles, restricted damage to starch and ended up producing superior quality oat bread (Hüttner et al., 2010). Henceforth it can be concluded that there is an obvious impact of refined rice flour particle size on the baking of gluten-free bread. On the other hand, further studies applying various kinds of gluten-free flours and various milling techniques are obligatory to confirm the above findings. From a nutritional viewpoint, the suitable combination could be incorporating a reduced volume and increased hardness of bread (De La Hera et al., 2014). Henceforth, apart to particle dimensions, hydrations of dough also have to be considered for modulating the hydrolysis of gluten-free breads and other similar foods.

### 8.8.2 Grinding and Air Classification

Once refined gluten-free flour is obtained, this could be exposed to various physical treatments to attain flours with diverse functionality and dietary formulation. The most exciting physical treatments are fine grinding (micronization) followed by air classification. This physical treatment involves decreasing the particle dimensions of flour significantly, that would alter flour functionality and formulate them highly appropriate for diverse processes.

Oat flours allow preparing breads of greater volume, reduced hardness and better sensory properties regards to breads baked by numerous other gluten-free flours (Hager et al., 2012). The superiority of oat breads may be enhanced if those flours are selected with better particle dimensions, decreased broken starch and protein percentage (Hager et al., 2012). The enzymatic alteration of their organic compounds could also be positive (Flander et al., 2011). Contrasting other gluten-free flours, oat flours possesses an elevated protein percentage, warranting studies for air-classification in future works. In this regards, oat cereal holding various protein percentage and particle dimensions could be accessed using fine grinding and air-classification, being the premium fractions that possess increased protein percentage (Wu & Stringfellow, 1995). β-glucan percentage in this part is also varied. Henceforth these portions should have a varying attribute in making of gluten-free bread, a phase that should be proceeded in detail as no present works regards to this exist. A numerous works on application of micronization and air classification in pea flours and legume flours have also been performed (Patel et al., 1980; Tyler et al., 1981). However, most of these works are chiefly based on the functional characteristics of these portions. Works on the addition of gluten-free flours to products are also very limited or void. On contrary, the integration of starches and protein is frequent in bread or cookie preparation, thus aforementioned flours can be an appealing alternative owing to their attribute of not involving artificial chemicals and ingredients.

### 8.8.3 Role of Hydrocolloids in Gluten-Free Breads

Additives for example hydrocolloids, enzymes and proteins are most commonly used in preparation of gluten-free bread with the objective of enhancing the visco-elastic attributes and end bread quality. Hydrocolloids are recently applied to enhance the rheological characteristics of gluten-free doughs and batters (Lazaridou et al., 2007), as they possess enormous prospective to structure three-dimensional polymer complex in solutions (Arendt et al., 2008). Various works have been conducted on the utilization of numerous hydrocolloids; say cellulose, guar gum, locust bean, hydroxypropril-methyl-cellulose and xanthan in gluten-free bread composition (Ahlborn et al., 2005; McCarthy et al., 2005; Moore et al., 2008). Respect to the other additives and refined flours used, particular hydrocolloids could affect to great percentage the bread volume and crumb texture of the baked bread, in which methyl celluloses was identified to be the most effectual amongst all (Lazaridou et al., 2007; Schober et al., 2007). Hydrocolloids are also significantly applied to enhance binding water capacity, dough viscosity, textural property, volume and end quality of bread (Mir et al., 2016). Methyl cellulose and xanthan gum are the widely applied hydrocolloids in gluten-free flour formulation owing to their their capability to advance the product quality (Hager & Arendt, 2013). Other hydrocolloids say CMC, guar gum, and locus bean gum are also most commonly applied in gluten-free bread dough preparation. However presently, numerous other hydrocolloids namely sodium carboxymethyl cellulose, cress see gum and (NaCMC) (Raeder et al., 2008) have also been recommended as novel gluten replacements that ensured promising baked bread quality. However, it was also observed that half-baked breads showed decreased volumes and increased crumb appearances, and elevated hardness. The incorporation of hydrocolloids, in peculiar CMC, partly mitigated the produced negative effect.

### 8.8.4 Prebiotic Gluten-Free Bread

The rising consumer insist for foodstuffs which are not only delicious and healthful but also offer health aspects have encouraged studies on prebiotics. The enormous application of inulin in the food industries is regards on its functional properties. Inulin is of enormous concern for the progress of healthy food products as it concomitantly communicates to an wide array of consumer necessities (Stephen et al., 2017). Inulin is the most commonly researched functional components in gluten-free breads (GFB) affecting constructively the sensorial and characteristics of final bread and prolonging the shelf life (Capriles & Arêas, 2013). Nevertheless, as the proteins present in gluten-free refined rice flours are usually incapable to hold gases at process of fermentation and baking henceforth, the enzymes were widely used to enhance the superiority of gluten-free breads by encouraging protein complex and elastic nature by protein cross-linking. The most widely employed prebiotics in gluten-free bread preparation is microbial transglutaminase – TG which aids in protein-connecting (Lee et al., 2005; Ziobro et al., 2016).

Prebiotics say oligofructose, resistant starch and inulin are the most commonly studied functional dietary fibers for gluten-free bread preparation.

As per (Capriles & Arêas, 2013), gluten-free bread with high percentage (4–12%) of inulin-type fructans (ITFs) exhibited specific volume less than 10%, whilst noticing a turn down higher than 10%. The authors recommended that ITFs could develop a gel arrangement and hold carbon dioxide similar to few hydrocolloids. Various degrees of polymerization of inulin also produce significant impact on the bread quality. Usually, reduced degree of polymerization of inulin has robust effects compared to superior ones (Ziobro et al., 2013). Resistant starch develops numerous functional properties and could not only lower the food energy but also improves digestive properties and final bread properties (Witczak et al., 2016). Furthermore, resistant starch does not affect the bread crumb firmness whereas enhances its rheological properties particularly , porosity and elasticity (Tsatsaragkou et al., 2014).

## 8.9 Conclusion

In answer to a global growing occurrence of celiac disease in individuals, the requirement to propose celiac disease patients with significant quality and extensive multiplicity of gluten-free baking food products is a plight. Nevertheless, the non-existence of gluten, whose existence decides the comprehensive appearance and textural attributes of bread making products, makes it a scientific dispute. This book chapter discusses numerous alternative resources, functional components (incorporated independently or in combination), and technologies that can produce gluten-free bread of enviable quality. Literature demonstrates that an imperative objective is to mimic the gluten-network by conjunction numerous components, from which hydrocolloids possess a decisive part. As well crosslinking enzymes have been progressively studied. In the future, additional research and investigations warrants to be focused on the detection and relevance of further novel gluten replacements and the development and popularization of the coeliac-safe wheat. Study on amalgamation of these outlooks must be performed to remark the impending synthetic impacts and produce gluten-free bread and other products with attributes resembling those of wheat breads. Conversely, elementary understanding about these baking substitutes on product superiority, consumer approval and shelf life has yet to be considered in more detail.

## References

Ahlborn, G. J., Pike, O. A., Hendrix, S. B., Hess, W. M., & Huber, C. S. (2005). Sensory, mechanical, and microscopic evaluation of staling in low-protein and gluten-free breads. *Cereal Chemistry, 82*(3), 328–335.

Alvarez-Jubete, L., Arendt, E., & Gallagher, E. (2009). Nutritive value and chemical composition of pseudocereals as gluten-free ingredient. *International Journal of Food Sciences and Nutrition, 60*(sup4), 240–257.

Aprodu, I., & Banu, I. (2015). Influence of dietary fiber, water, and glucose oxidase on rheological and baking properties of maize based gluten-free bread. *Food Science and Biotechnology, 24*(4), 1301–1307.

Araki, E., Ikeda, T. M., Ashida, K., Takata, K., Yanaka, M., & Iida, S. (2009). Effects of rice flour properties on specific loaf volume of one-loaf bread made from rice flour with wheat vital gluten. *Food Science and Technology Research, 15*(4), 439–448.

Arendt, E. K., Ryan, L. A., & Dal Bello, F. (2007). Impact of sourdough on the texture of bread. *Food Microbiology, 24*(2), 165–174.

Arendt, E. K., Morrissey, A., Moore, M. M., & Dal Bello, F. (2008). Gluten-free breads. In *Gluten-free cereal products and beverages* (p. 289). Elsevier.

Basso, F. M., Mangolim, C. S., Aguiar, M. F. A., Monteiro, A. R. G., Peralta, R. M., & Matioli, G. (2015). Potential use of cyclodextrin-glycosyltransferase enzyme in bread-making and the development of gluten-free breads with pinion and corn flours. *International Journal of Food Sciences and Nutrition, 66*(3), 275–281.

Campbell, G. M., & Martin, P. (2012). Bread aeration and dough rheology: An introduction. In *Breadmaking* (pp. 299–336). Elsevier.

Campbell, G. M., & Mougeot, E. (1999). Creation and characterisation of aerated food products. *Trends in Food Science & Technology, 10*(9), 283–296.

Cappa, C., Lucisano, M., Raineri, A., Fongaro, L., Foschino, R., & Mariotti, M. (2016). Gluten-free bread: Influence of sourdough and compressed yeast on proofing and baking properties. *Foods, 5*(4), 69.

Capriles, V. D., & Arêas, J. A. (2013). Effects of prebiotic inulin-type fructans on structure, quality, sensory acceptance and glycemic response of gluten-free breads. *Food & Function, 4*(1), 104–110.

Cauvain, S. P., & Young, L. S. (2006). *The Chorleywood bread process*. Woodhead Publishing.

Cheng, L.-M. (1992). *Food machinery: For the production of cereal foods, snack foods and confectionery*. Elsevier.

Chin, N. L., & Campbell, G. M. (2005). Dough aeration and rheology: Part 2. Effects of flour type, mixing speed and total work input on aeration and rheology of bread dough. *Journal of the Science of Food and Agriculture, 85*(13), 2194–2202.

Chin, L. N., Tan, C. M., Pa, N. F. C., & Yusof, Y. A. (2015). *Method and apparatus for high intensity ultrasonic treatment of baking materials*. Google Patents.

Chiotellis, E., & Campbell, G. M. (2003). Proving of bread dough I: Modelling the evolution of the bubble size distribution. *Food and Bioproducts Processing, 81*(3), 194–206.

de la Hera, E., Gomez, M., & Rosell, C. M. (2013a). Particle size distribution of rice flour affecting the starch enzymatic hydrolysis and hydration properties. *Carbohydrate Polymers, 98*(1), 421–427.

de la Hera, E., Martinez, M., & Gómez, M. (2013b). Influence of flour particle size on quality of gluten-free rice bread. *LWT- Food Science and Technology, 54*(1), 199–206.

de la Hera, E., Talegón, M., Caballero, P., & Gómez, M. (2013c). Influence of maize flour particle size on gluten-free breadmaking. *Journal of the Science of Food and Agriculture, 93*(4), 924–932.

De La Hera, E., Rosell, C. M., & Gomez, M. (2014). Effect of water content and flour particle size on gluten-free bread quality and digestibility. *Food Chemistry, 151*, 526–531.

De Vuyst, L., & Vancanneyt, M. (2007). Biodiversity and identification of sourdough lactic acid bacteria. *Food Microbiology, 24*(2), 120–127.

Demirkesen, I., Mert, B., Sumnu, G., & Sahin, S. (2010). Utilization of chestnut flour in gluten-free bread formulations. *Journal of Food Engineering, 101*(3), 329–336.

Deora, N. S., Deswal, A., Dwivedi, M., & Mishra, H. N. (2014). Prevalence of coeliac disease in India: A mini review. *International Journal of Latest Research Science and Technology, 3*(10), 58–60.

Flander, L., Holopainen, U., Kruus, K., & Buchert, J. (2011). Effects of tyrosinase and laccase on oat proteins and quality parameters of gluten-free oat breads. *Journal of Agricultural and Food Chemistry, 59*(15), 8385–8390.

Gänzle, M. G., Loponen, J., & Gobbetti, M. (2008). Proteolysis in sourdough fermentations: Mechanisms and potential for improved bread quality. *Trends in Food Science & Technology, 19*(10), 513–521.

Gunness, P., & Gidley, M. J. (2010). Mechanisms underlying the cholesterol-lowering properties of soluble dietary fibre polysaccharides. *Food & Function, 1*(2), 149–155.

Hager, A.-S., & Arendt, E. K. (2013). Influence of hydroxypropylmethylcellulose (HPMC), xanthan gum and their combination on loaf specific volume, crumb hardness and crumb grain characteristics of gluten-free breads based on rice, maize, teff and buckwheat. *Food Hydrocolloids, 32*(1), 195–203.

Hager, A.-S., Wolter, A., Czerny, M., Bez, J., Zannini, E., Arendt, E. K., & Czerny, M. (2012). Investigation of product quality, sensory profile and ultrastructure of breads made from a

range of commercial gluten-free flours compared to their wheat counterparts. *European Food Research and Technology, 235*(2), 333–344.
Hicsasmaz, Z., Yazgan, Y., Bozoglu, F., & Katnas, Z. (2003). Effect of polydextrose-substitution on the cell structure of the high-ratio cake system. *LWT- Food Science and Technology, 36*(4), 441–450.
Houben, A., Götz, H., Mitzscherling, M., & Becker, T. (2010). Modification of the rheological behavior of amaranth (Amaranthus hypochondriacus) dough. *Journal of Cereal Science, 51*(3), 350–356.
Hüttner, E. K., Dal Bello, F., & Arendt, E. K. (2010). Rheological properties and bread making performance of commercial wholegrain oat flours. *Journal of Cereal Science, 52*(1), 65–71.
Jekle, M., & Becker, T. (2011). Dough microstructure: Novel analysis by quantification using confocal laser scanning microscopy. *Food Research International, 44*(4), 984–991.
Jekle, M., Houben, A., Mitzscherling, M., & Becker, T. (2010). Effects of selected lactic acid bacteria on the characteristics of amaranth sourdough. *Journal of the Science of Food and Agriculture, 90*(13), 2326–2332.
Jerome, R. E., Singh, S. K., & Dwivedi, M. (2019). Process analytical technology for bakery industry: A review. *Journal of Food Process Engineering, 42*(5), e13143.
Kadan, R., Bryant, R., & Miller, J. (2008). Effects of milling on functional properties of rice flour. *Journal of Food Science, 73*(4), E151–E154.
Kagnoff, M. F. (2005). Overview and pathogenesis of celiac disease. *Gastroenterology, 128*(4), S10–S18.
Lazaridou, A., Duta, D., Papageorgiou, M., Belc, N., & Biliaderis, C. G. (2007). Effects of hydrocolloids on dough rheology and bread quality parameters in gluten-free formulations. *Journal of Food Engineering, 79*(3), 1033–1047.
Lee, S., Kim, S., & Inglett, G. E. (2005). Effect of shortening replacement with oatrim on the physical and rheological properties of cakes. *Cereal Chemistry, 82*(2), 120–124.
López-Tenorio, J. A., Rodríguez-Sandoval, E., & Sepúlveda-Valencia, J. U. (2015). The influence of different emulsifiers on the physical and textural characteristics of gluten-free cheese bread. *Journal of Texture Studies, 46*(4), 227–239.
Massey, A., Khare, A., & Niranjan, K. (2001). Air inclusion into a model cake batter using a pressure whisk: Development of gas hold-up and bubble size distribution. *Journal of Food Science, 66*(8), 1152–1157.
McCarthy, D., Gallagher, E., Gormley, T., Schober, T., & Arendt, E. (2005). Application of response surface methodology in the development of gluten-free bread. *Cereal Chemistry, 82*(5), 609–615.
Mir, S. A., Shah, M. A., Naik, H. R., & Zargar, I. A. (2016). Influence of hydrocolloids on dough handling and technological properties of gluten-free breads. *Trends in Food Science & Technology, 51*, 49–57.
Mishra, N., Tripathi, R., & Dwivedi, M. (2020). Development and characterization of antioxidant rich wheatgrass cupcake. *Carpathian Journal of Food Science & Technology, 12*(3).
Mohammadi, M., Sadeghnia, N., Azizi, M.-H., Neyestani, T.-R., & Mortazavian, A. M. (2014). Development of gluten-free flat bread using hydrocolloids: Xanthan and CMC. *Journal of Industrial and Engineering Chemistry, 20*(4), 1812–1818.
Moore, M., Juga, B., Schober, T., & Arendt, E. (2007). Effect of lactic acid bacteria on properties of gluten-free sourdoughs, batters, and quality and ultrastructure of gluten-free bread. *Cereal Chemistry, 84*(4), 357–364.
Moore, M. M., Dal Bello, F., & Arendt, E. K. (2008). Sourdough fermented by Lactobacillus plantarum FSTá1. 7 improves the quality and shelf life of gluten-free bread. *European Food Research and Technology, 226*(6), 1309–1316.
Moroni, A. V., Dal Bello, F., & Arendt, E. K. (2009). Sourdough in gluten-free bread-making: An ancient technology to solve a novel issue? *Food Microbiology, 26*(7), 676–684.

Moroni, A. V., Arendt, E. K., Morrissey, J. P., & Dal Bello, F. (2010). Development of buckwheat and teff sourdoughs with the use of commercial starters. *International Journal of Food Microbiology, 142*(1–2), 142–148.

Mustalahti, K., Catassi, C., Reunanen, A., Fabiani, E., Heier, M., McMillan, S., Murray, L., Metzger, M.-H., Gasparin, M., & Bravi, E. (2010). The prevalence of celiac disease in Europe: Results of a centralized, international mass screening project. *Annals of Medicine, 42*(8), 587–595.

Niewinski, M. M. (2008). Advances in celiac disease and gluten-free diet. *Journal of the American Dietetic Association, 108*(4), 661–672.

Park, J. H., Kim, D. C., Lee, S. E., Kim, O. W., Kim, H., Lim, S. T., & Kim, S. S. (2014). Effects of rice flour size fractions on gluten free rice bread. *Food Science and Biotechnology, 23*(6), 1875–1883.

Patel, K., Bedford, C., & Youngs, C. (1980). Navy bean flour fractions. *Cereal Chemistry, 57*(2), 123–125.

Raeder, J., Larson, D., Li, W., Kepko, E. L., & Fuller-Rowell, T. (2008). OpenGGCM simulations for the THEMIS mission. *Space Science Reviews, 141*(1-4), 535–555.

Richardson, G., Langton, M., Fäldt, P., & Hermansson, A. M. (2002). Microstructure of α-crystalline emulsifiers and their influence on air incorporation in cake batter. *Cereal Chemistry, 79*(4), 546–552.

Rifna, E. J., Singh, S. K., Chakraborty, S., & Dwivedi, M. (2019). Effect of thermal and non-thermal techniques for microbial safety in food powder: Recent advances. *Food Research International, 126*, 108654.

Schober, T. J., Bean, S. R., & Boyle, D. L. (2007). Gluten-free sorghum bread improved by sourdough fermentation: Biochemical, rheological, and microstructural background. *Journal of Agricultural and Food Chemistry, 55*(13), 5137–5146.

Segura, M. E. M., & Rosell, C. M. (2011). Chemical composition and starch digestibility of different gluten-free breads. *Plant Foods for Human Nutrition, 66*(3), 224.

Sinelli, N., Casiraghi, E., & Downey, G. (2008). Studies on proofing of yeasted bread dough using near-and mid-infrared spectroscopy. *Journal of Agricultural and Food Chemistry, 56*(3), 922–931.

Stefańska, I., Piasecka-Jóźwiak, K., Kotyrba, D., Kolenda, M., & Stecka, K. M. (2016). Selection of lactic acid bacteria strains for the hydrolysis of allergenic proteins of wheat flour. *Journal of the Science of Food and Agriculture, 96*(11), 3897–3905.

Stephen, A. M., Champ, M. M.-J., Cloran, S. J., Fleith, M., van Lieshout, L., Mejborn, H., & Burley, V. J. (2017). Dietary fibre in Europe: Current state of knowledge on definitions, sources, recommendations, intakes and relationships to health. *Nutrition Research Reviews, 30*(2), 149–190.

Svensson, L., Sekwati-Monang, B., Lutz, D. L., Schieber, A., & Ganzle, M. G. (2010). Phenolic acids and flavonoids in nonfermented and fermented red sorghum (Sorghum bicolor (L.) Moench). *Journal of Agricultural and Food Chemistry, 58*(16), 9214–9220.

Trinh, L., Lowe, T., Campbell, G., Withers, P., & Martin, P. (2013). Bread dough aeration dynamics during pressure step-change mixing: Studies by X-ray tomography, dough density and population balance modelling. *Chemical Engineering Science, 101*, 470–477.

Tripathi, R., Sharma, D., Dwivedi, M., Rizvi, S. I., & Mishra, N. (2017). Wheatgrass incorporation as a viable strategy to enhance nutritional quality of an edible formulation. *Annals of Phytomedicine, 6*(1), 68–75.

Tsatsaragkou, K., Gounaropoulos, G., & Mandala, I. (2014). Development of gluten free bread containing carob flour and resistant starch. *LWT- Food Science and Technology, 58*(1), 124–129.

Tyler, R., Youngs, C., & Sosulski, F. (1981). Composition of the starch and protein fractions. *Cereal Chemisty, 58*(2), 144–148.

Utrilla-Coello, R., Bello-Perez, L. A., Vernon-Carter, E., Rodriguez, E., & Alvarez-Ramirez, J. (2013). Microstructure of retrograded starch: Quantification from lacunarity analysis of SEM micrographs. *Journal of Food Engineering, 116*(4), 775–781.

Wang, K., Lu, F., Li, Z., Zhao, L., & Han, C. (2017). Recent developments in gluten-free bread baking approaches: A review. *Food Science and Technology, 37*, 1–9.
Witczak, M., Ziobro, R., Juszczak, L., & Korus, J. (2016). Starch and starch derivatives in gluten-free systems–A review. *Journal of Cereal Science, 67*, 46–57.
Wronkowska, M., Haros, M., & Soral-Śmietana, M. (2013). Effect of starch substitution by buckwheat flour on gluten-free bread quality. *Food and Bioprocess Technology, 6*(7), 1820–1827.
Wu, Y. V., & Stringfellow, A. C. (1995). *Enriched protein-and beta-glucan fractions from high-protein oats by air classification*.
Ziobro, R., Korus, J., Juszczak, L., & Witczak, T. (2013). Influence of inulin on physical characteristics and staling rate of gluten-free bread. *Journal of Food Engineering, 116*(1), 21–27.
Ziobro, R., Juszczak, L., Witczak, M., & Korus, J. (2016). Non-gluten proteins as structure forming agents in gluten free bread. *Journal of Food Science and Technology, 53*(1), 571–580.

# Chapter 9
# Overview of the Gluten-Free Market

**Rewa Kulshrestha**

**Abstract** Recent trends and consumer insights suggest that the market for gluten free products will continue to rise globally. The technological advancement with reference to the gluten detection method will help to create the global awareness of the celiac disease and this would bring the radical change in the market as well the rise of demand among the consumers. This chapter aims to disucss the insight of gluten free market.

**Keywords** Gluten-Free Market · Bakery · Nutrition

## 9.1 Introduction

As per latest insight, various reports suggest that the global gluten-free products market size is estimated for **USD 5.7 billion in 2020** and is projected to reach USD 8.3 billion by 2025 at the rate of CAGR of 8.1% ( CAGR). The market is driven by the mounting prevalence of celiac disease and trends in towards health. Over the last decades, it has been witnessed that globally companies in the area of gluten free sector are experiencing greater demand from consumers with intolerance towards gluten. This trend will definitely continue to rise and overall rapidly move up the market.

R. Kulshrestha (✉)
Department of Food Processing and Technology, Atal Bihari Vajpayee Vishwavidyalaya, Bilaspur, Chhattisgarh, India
e-mail: rewakulshrestha@bilaspuruniversity.ac.in



N. Singh Deora et al. (eds.), *Challenges and Potential Solutions in Gluten Free Product Development*, Food Engineering Series,
https://doi.org/10.1007/978-3-030-88697-4_9

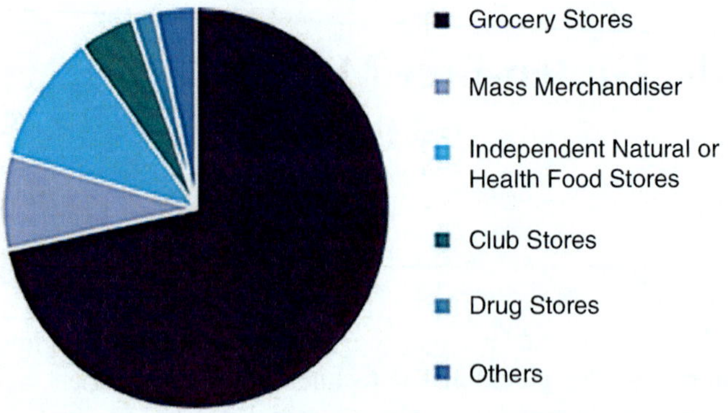


**Figure Global Gluten Free Products Market Share by distribution Channel, 2019 (%). (Source : www.grandviewresearch.com)**

## 9.2 Impact of COVID-19 on the Current Market Size and Forecast

The COVID-19 pandemic is expected to bring some negative impact on the global impact primarily due to importing and exporting the end products, owing to the limited international trading activities (Miranda et al. 2014). However, we are optimistic that post-pandemic, the market for gluten-free products is projected to witness an upward growth trend, owing to the rise in consumers preferring gluten-free and other free-from foods, as a result of a shift in the consumption lifestyles towards healthier eating (Table 9.1).

## 9.3 Market Dynamics – Drivers and Opportunities

With the advent of better and more reliable diagnostic methods, we are able to witness and capture celiac disease data. This data captures the prevalence of celiac disease worldwide which will in turn boost the market of gluten free products in coming decade. Thus, in terms of drivers, more simple and reliable diagnostics methods will create more opportunities in the space of gluten free product development and global outreach. Another reason for upwards market growth will also be due to increased incidence of non-celiac gluten sensitivity as well as other related inflammatory diseases, and autoimmune disorders. This is supported by active participation of various government campaigns worldwide.

**Table 9.1** List of global companies in gluten free space

| |
|---|
| 1. The Kraft Heinz Company (US), |
| 2. The Hain Celestial Group Inc (US), |
| 3. General Mills (US), |
| 4. Kellogg's Company (US), |
| 5. ConAgra Brands Inc (US), |
| 6. Hero AG (Switzerland), |
| 7. Barilla G.E.R Fratelli S.P.A (Italy), |
| 8. Quinoa Corporation (US), |
| 9. Freedom Foods Group Limited (Australia) |
| 10. Koninklijke Wessanen N.V (Netherlands) |
| 11. Raisio PLC (Finland), |
| 12. Dr Schär AG/SPA (Italy) |
| 13. Enjoy Life Foods (US) |
| 14. Farmo S.P.A. (Italy) |
| 15. Big OZ (UK) |
| 16. Alara Wholefoods Ltd (UK) |
| 17. Norside Foods Ltd (UK) |
| 18. Warburtons (UK) |
| 19. Silly Yaks (Australia) |
| 20. Seitz Glutenfrei GMBH (Germany) |
| 21. Bob's Red Mill (US) |
| 22. Kelkin Ltd (Ireland) |
| 23. Amy's Foods (US) |
| 24. Golden West Specialty Foods (US) |
| 25. Prima Foods (UK) |
| 26. Katz Gluten Free (US) |
| 27. Genius Foods (UK) |
| 28. Chosen Foods LLC (US) |
| 29. BFree (Ireland) |
| 30. Mickey's LLC (US) |
| 31. Rachel Pauls Food (US) |
| 32. Gee Free LLC (US) |
| 33. Fody Foods (Canada) |
| 34. Gluten-free Prairie (US) |
| 35. Gluten Free Cornwall (UK) |
| 36. Feel Good Foods (US) |
| 37. Canyon Bakehouse LLC (US) |
| 38. Barr Necessities (US) |
| 39. Avena Foods Limited (Canada) |
| 40. Complete Start (US) |

## 9.4 Challenges

If we globally analyze the nutritional information about the gluten free products, one of the concern is the lack of adequate fibers in the products (Gallagher, 2009; Rosell & Matos, 2015). This ultimately leads to the alignments of the digestive system. We need Gluten-free products lacks an adequate amount of dietary fibers, resulting in constipation and other ailments of digestive system. We need to address this challenge in the near future to further uplift the market of gluten free products globally.

In the article, we have attempted to present the data for the gluten free Bread and Pasta which currently holds the prime categories for the growth of gluten free market globally.

## 9.5 Bakery Market – Gluten Free

Bakery products as per current market dynamics has one of the is largest share for the year 2020. Figure below presents the market dynamics across the global for the bread and bread related products in the retail market. With the rise in global celic disease, in proportion the market for the gluten free bread will also witness upward trends since customers would look for alternative to wheat based bread and bakery products (Foschia et al. 2016; Amanda topper 2017; Grandview research 2020) (Figs. 9.1 and 9.2 and Tables 9.2 and 9.3).

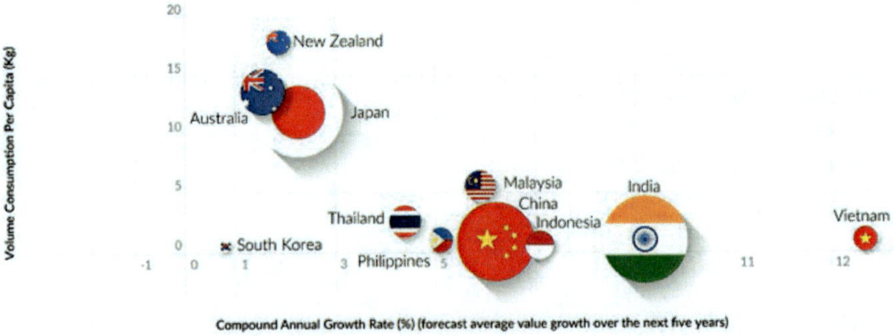


**Fig. 9.1** Asia Pacific (APAC) Retail Market Overview: Bread & Bread Products, 2020

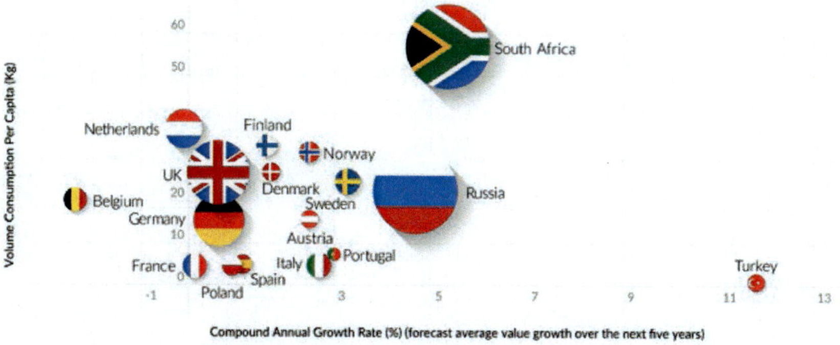


**Fig. 9.2** EMEA retail market overview: bread & bread products, 2020
Note: Market figures for 2020 onwards are estimations created prior to the global COVID-19 outbreak Base: bubble size is based on market volume (kg); CAGR is based on market growth in value over the last five years in local currency
Source: Mintel Market Sizes

## 9.6 Gluten-Free Pasta Market Outlook – 2025

Apart from the bread segment, gluten free pasta market is also rapidly growing (refer to figure below) . Current estimate suggest that the market was around $909.8 million in 2017 and by 2025 it is projected to reach $1,289.2 million with the impressive CAGR of 4.5%. Various key food manufacturers strategize on coming up with free-from food products that cater to the requirement of food intolerant consumers owing to the growth in concern of the consumers about the ingredients in the food. Gluten-free pasta is one of these which eventually triggers gluten-free pasta market growth in terms of value sales (Fig. 9.3).

## 9.7 Conclusions

In conclusion, the market for gluten free products will continue to rise. With recent advancement in science and technology and better methods to detect gluten intolerance, the market would continue to expand. Bakery and pasta segment would continue to lead the space of gluten free market. Innovation around bakery and pasta would drive the market to greater heights. However, we would need better and nutritious products that is designed by keeping consumer's requirement in mind.

**Table 9.2** List of gluten free bread along with countries

| Sr. | Product | Company | Market | Price | g |
|---|---|---|---|---|---|
| 1 | Gluten-Free Masterbaker's Classic White Bread Slices | Dr. Schär | Germany | 3.48 | 350 |
| 2 | Gluten-Free Masterbaker's Vital Grain Bread | Dr. Schär | Germany | 3.48 | 350 |
| 3 | Rice Almond Gluten Free Bread | Food For Life Baking | Panama | 9.5 | 680 |
| 4 | Gluten-Free Cereal Bread with Sourdough, Millet and Quinoa | Dr. Schär | Italy | 4.61 | 330 |
| 5 | Gluten-Free Rustic Soft Bread Slices | NT Food | Italy | 3.15 | 165 |
| 6 | Gluten-Free American Sandwich Bread | NT Food | Italy | 4.66 | 240 |
| 7 | Gluten-Free Classic Bread with Sourdough, Millet and Quinoa | Dr. Schär | Italy | 4.07 | 330 |
| 8 | Gluten-Free Bread Mix | Dr. Schär | Slovenia | 5.21 | 1000 |
| 9 | Gluten-Free Toasted Bread Slices | NT Food | Italy | 3.74 | 225 |
| 10 | Gluten-Free Cereal Bread with Leaven, Millet and Quinoa | Dr. Schär | France | 4.09 | |
| 11 | Gluten-Free Focaccia Bread with Rosemary | Dr. Schär | Norway | 5.28 | 200 |
| 12 | Gluten-Free Maestro Vital Bread | Dr. Schär | Czech Republic | 3.86 | 350 |
| 13 | Mix B Gluten-Free Bread-Mix | Dr. Schär | Slovenia | 5.78 | 1000 |
| 14 | Gluten-Free Cereal Bread | Dr. Schär | Czech Republic | 2.31 | 300 |
| 15 | Gluten-Free Bread Mix | Dr. Schär | Spain | 4.5 | 1000 |
| 16 | Organic Gluten-Free Focaccia Bread | Schnitzer | Netherlands | 5.2 | 110 |
| 17 | Organic Gluten-Free Rustic Bread with Amaranth | Schnitzer | Netherlands | 6.98 | 500 |
| 18 | Gluten-Free Cereal Bread | Dr. Schär | Italy | 4.22 | 300 |
| 19 | Gluten-Free Focaccia Bread with Rosemary | Dr. Schär | Germany | 3.47 | 200 |
| 20 | Gluten-Free Rustic Sourdough Bread Slices | Conad | Italy | 2.77 | 300 |
| 21 | Gluten-Free Toasted Bread Slices | Dr. Schär | Italy | 5.64 | 86.7 |
| 22 | Gluten Free White Soft Bread Slices | NT Food | Italy | 3.38 | 300 |
| 23 | Gluten Free French Loaf Artisan Bread Mix | Williams-Sonoma | USA | 14.95 | 459.27 |
| 24 | Organic Gluten-Free Cereal Panini Bread | Dr. Schär | Spain | 4.49 | 165 |
| 25 | Triple Seeded Farmhouse Gluten-Free Bread | Genius Foods | Netherlands | 4.39 | 535 |
| 26 | Gluten Free Baker's Bread | Dr. Schär | Germany | 3.8 | 300 |
| 27 | Gluten Free Bread Mix | NT Food | Italy | 6.6 | 1000 |

(continued)

**Table 9.2** (continued)

| Sr. | Product | Company | Market | Price | g |
|---|---|---|---|---|---|
| 28 | Gluten Free Master Baker's Multigrain Bread | Dr. Schär | Germany | 3.18 | 300 |
| 29 | Gluten-Free Multigrain Buckwheat Bread | Dr. Schär | Poland | 2.72 | 250 |
| 30 | Homemade Gluten Free Coffee Cake Quick Bread Mix | Williams-Sonoma | USA | 16.95 | 538.65 |
| 31 | Gluten-Free Bread B Mix | Dr. Schär | Italy | 2.14 | 500 |
| 32 | Gluten Free Mix B Bread-Mix | Dr. Schär | Hungary | 4.91 | 1000 |
| 33 | Gluten-Free Wholewheat Bread Mix | NT Food | Italy | 5.67 | 1000 |
| 34 | Gluten-Free Toasted Bread Slices with Cereals | Dr. Schär | Italy | 5.2 | 86.7 |
| 35 | Gluten-Free Sourdough Bread with Cranberries and Lingonberries | Dr. Schär | Germany | 3.8 | 240 |
| 36 | Gluten-Free Toasted Bread | Dr. Schär | France | 3.37 | 150 |
| 37 | Gluten Free White Soft Bread Slices | NT Food | Italy | 2.51 | 165 |
| 38 | Gluten-Free Soft Seeded Sliced Bread | Dr. Schär | Slovakia | 2.51 | 250 |
| 39 | Gluten Free Mix B Bread-Mix | Dr. Schär | Denmark | 7.08 | 1000 |
| 40 | Gluten Free Buckwheat Bread | Dr. Schär | Serbia | 5.17 | 240 |
| 41 | Gluten-Free Spiced Apple Quick Bread Mix | Williams-Sonoma | USA | 17.95 | 510.3 |
| 42 | Gluten Free Crostini Bread | Dr. Schär | Israel | 8.23 | 150 |
| 43 | Gluten-Free Multi-Grain Vital Bread | Dr. Schär | Netherlands | 3.69 | 350 |
| 44 | Gluten-Free Bread with Raisins | Food For Life Baking | Mexico | 10.78 | 680 |
| 45 | Gluten Free Soft Seeded Sliced Bread | Dr. Schär | Morocco | 3.02 | 250 |
| 46 | Organic Gluten-Free Rustic Bread with Amaranth | Schnitzer | Germany | 5 | 500 |
| 47 | Gluten-Free Crostini Bread | Dr. Schär | Italy | 3.37 | 150 |
| 48 | Master Baker's Gluten-Free Multigrain Sliced Bread | Dr. Schär | Switzerland | 4.72 | 300 |
| 49 | Gluten Free Classic White Sliced Bread | Dr. Schär | Switzerland | 4.72 | 300 |
| 50 | Gluten-Free Whole meal Bread Mix | NT Food | Czech Republic | 5.29 | 1000 |
| 51 | Gluten Free Wholegrain Oat Bread Mix | Bauck | Spain | 3.95 | 500 |
| 52 | Organic Gluten-Free Quick Bread Mix | Bauck | Spain | 4.52 | 500 |
| 53 | Gluten Free Black Bread Mix | Bauck | Spain | 3.95 | 500 |
| 54 | Gluten-Free Dark Sourdough Bread | Dr. Schär | Denmark | 4.53 | 350 |
| 55 | Gluten-Free Country Bread | Dr. Schär | Germany | 3.31 | 275 |
| 56 | Gluten-Free Panbauletto Sliced Bread | NT Food | Czech Republic | 3.5 | 300 |
| 57 | Gluten Free Mix B Bread Mix | Dr. Schär | Netherlands | 2.54 | 500 |

(continued)

**Table 9.2** (continued)

| Sr. | Product | Company | Market | Price | g |
|---|---|---|---|---|---|
| 58 | Gluten-Free Light Bread Rolls | Genius Foods | Germany | 4.51 | 320 |
| 59 | Gluten-Free Hamburger Bread | Dr. Schär | Morocco | 5.3 | 75 |
| 60 | Gluten Free Sliced Sesame Bread | Schnitzer | Switzerland | 4.73 | 250 |
| 61 | Gluten-free Homemade Bread | Dr. Schär | Brazil | 5.14 | 240 |
| 62 | Gluten Free Sandwich Bread with Seeds and Chia | Dr. Schär | Norway | 5.01 | 400 |
| 63 | Gluten Free Dark Sandwich Bread with Sourdough | Dr. Schär | Norway | 5.35 | 400 |
| 64 | Gluten-Free Pangette Sliced Bread | NT Food | Czech Republic | 4.46 | 340 |
| 65 | Gluten Free Buckwheat Bread | Dr. Schär | Turkey | 4.22 | 240 |
| 66 | Gluten-Free Ciabatta Bread | Dr. Schär | Morocco | 5.18 | 50 |
| 67 | Gluten Free Black Bread | Bauck | Poland | 4.1 | 500 |
| 68 | Gluten-Free Country Bread | Dr. Schär | Czech Republic | 4.42 | 240 |
| 69 | Gluten Free Country Bread | Dr. Schär | Croatia | 4.44 | 240 |
| 70 | Homemade Gluten Free Spiced Pecan Pumpkin Bread Mix | Williams-Sonoma | USA | 16.95 | 510.3 |
| 71 | Gluten Free White Bread | Genius Foods | Germany | 4.38 | 400 |
| 72 | Homemade Gluten Free Pumpkin Cheesecake Quick Bread Mix | Williams-Sonoma | USA | 17.95 | 510.3 |
| 73 | Gluten-Free Panbauletto Sliced Bread | NT Food | Czech Republic | 4.41 | 350 |
| 74 | Gluten-Free Sliced Cereal Bread | Genius Foods | France | 5.28 | 400 |
| 75 | Gluten-Free Homemade Bread Slices | NT FOOD | Italy | 11.63 | 300 |
| 76 | Gluten-Free Wholemeal Bread Slices | NT FOOD | Italy | 11.63 | 340 |
| 77 | Gluten-Free Brown Sliced Bread | Dr. Schär | Slovakia | 2.32 | 250 |
| 78 | Gluten-Free Multigrain Country Bread | Dr. Schär | Germany | 3.49 | 250 |
| 79 | Gluten Free Country Bread | Dr. Schär | Switzerland | 4.21 | 240 |
| 80 | Gluten Free Country Bread | Dr. Schär | Germany | 3.44 | 240 |
| 81 | Gluten-Free Wholegrain Oat Bread Mix | Bauck | Germany | 4.46 | 500 |
| 82 | Gluten Free Quick Bread Mix with Seeds | Bauck | Germany | 4.46 | 500 |
| 83 | Gluten-Free B-Bread Mix | Dr. Schär | Switzerland | 5.96 | 1000 |
| 84 | Gluten-Free Homemade-Style Bread Slices | Dr. Schär | Italy | 4.42 | 240 |
| 85 | Gluten-Free Bread Flour Mix B | Dr. Schär | Italy | 5.97 | 1020 |
| 86 | Gluten-Free Bread Sticks with Buckwheat | Dr. Schär | Spain | 1.54 | 15 |
| 87 | Gluten Free Country Bread | Dr. Schär | Denmark | 6.4 | 240 |
| 88 | Gluten-Free Sandwich Bread | Dr. Schär | Austria | 4.33 | 400 |

(continued)

**Table 9.2** (continued)

| Sr. | Product | Company | Market | Price | g |
|---|---|---|---|---|---|
| 89 | Gluten-Free Bread Sticks with Spreadable Cocoa Cream | Dr. Schär | Spain | 2.86 | 52 |
| 90 | Gluten-Free Black Bread Mix | Bauck | Germany | 4.56 | 500 |
| 91 | Gluten Free White Bread | Dr. Schär | Mexico | 5.7 | 300 |
| 92 | Gluten-Free White Bread | Dr. Schär | Morocco | 2.18 | 200 |
| 93 | Gluten Free Ciabatta Bread | Dr. Schär | Chile | 5.54 | 50 |
| 94 | Gluten Free Mix B Bread Mix | Dr. Schär | Netherlands | 5.22 | 1000 |
| 95 | Gluten Free Granary Sandwich Bread with Chia | Dr. Schär | Germany | 4.36 | 400 |
| 96 | Gluten-Free Kornspitz Bread Rolls | Dr. Schär | Austria | 2.74 | 70 |
| 97 | Gluten-Free Wholewheat Panfette Sliced Bread | NT Food | Croatia | 6.62 | 85 |
| 98 | Gluten-Free Maple Pecan Quick Bread Mix | Williams-Sonoma | USA | 16.95 | 510.3 |
| 99 | Organic Gluten Free Sunflower Bread | REMA 1000 | Norway | 3.67 | 250 |
| 100 | Organic Gluten Free Maize Bread with Amaranth | REMA 1000 | Norway | 3.79 | 250 |
| 101 | Gluten Free Multigrain Country Bread | Dr. Schär | Germany | 3.38 | 250 |
| 102 | Mix B Gluten Free Bread Mix | Dr. Schär | Hungary | 5.48 | 1000 |
| 103 | Gluten-Free Sweet Bread Rolls with Chocolate Drops | Dr. Schär | Italy | 3.92 | 55 |
| 104 | Gluten Free Sliced Bread | Dr. Schär | Switzerland | 4.95 | 350 |
| 105 | Gluten-Free Panfette Sliced Bread | NT Food | Czech Republic | 4.06 | 75 |
| 106 | Gluten-Free Rosette Bread Rolls | Dr. Schär | Italy | 4.79 | 58 |
| 107 | Gluten-Free Bread with Sesame | Schnitzer | Costa Rica | 5.2 | 250 |
| 108 | Gluten-Free Rustica Bread + Fiber | Dr. Schär | Turkey | 4.69 | 225 |
| 109 | Gluten-Free Homemade Bread | Dr. Schär | Turkey | 6.93 | 240 |
| 110 | Gluten-Free Sliced White Bread | Dr. Schär | Turkey | 6.1 | 200 |
| 111 | Gluten-Free Meyer Lemon Quick Bread | Williams-Sonoma | USA | 14.95 | 510.3 |
| 112 | Gluten Free Focaccia Bread with Rosemary | Dr. Schär | Germany | 3.17 | 66 |
| 113 | Gluten-Free Bread Sticks with Cocoa Cream | Dr. Schär | Poland | 1.35 | 52 |
| 114 | Gluten Free Bread Mix | Dr. Schär | Croatia | 5.67 | 1000 |
| 115 | Gluten-Free Sliced Bread with Raisins | Genius Foods | France | 4.59 | 400 |
| 116 | Gluten-Free Hamburger Bread Rolls | NT FOOD | France | 3.2 | 180 |
| 117 | Gluten-Free Multigrain Bread | Dr. Schär | Norway | 4.91 | 300 |
| 118 | Maestro Classic Gluten-Free Bread | Dr. Schär | France | 3.7 | 300 |
| 119 | Gluten Free Whole Wheat Bread | Dr. Schär | Mexico | 7.95 | 350 |
| 120 | Gluten Free Classic Sliced Bread | Genius Foods | Portugal | 4.69 | 400 |
| 121 | Gluten Free Multigrain Bread | Dr. Schär | Denmark | 5.71 | 300 |

(continued)

**Table 9.2** (continued)

| Sr. | Product | Company | Market | Price | g |
|---|---|---|---|---|---|
| 122 | Gluten-Free Dark Sliced Bread | Dr. Schär | Finland | 3.54 | 350 |
| 123 | Gluten Free Sweet Cinnamon Apple Quick Bread Mix | Williams-Sonoma | USA | 12.95 | 498.96 |
| 124 | Gluten-Free Wholegrain Oat Bread Mix | Bauck | Denmark | 5.84 | 500 |
| 125 | Gluten Free Bread Mix | Dr. Schär | Denmark | 7.81 | 1000 |
| 126 | Gluten Free Dark Bread Mix | Dr. Schär | Denmark | 9.06 | 1000 |
| 127 | Gluten-Free Hamburger Bread | Dr. Schär | Italy | 3.81 | 300 |
| 128 | Gluten-free Cereal Bread | Dr. Schär | Spain | 2.99 | 300 |
| 129 | Gluten Free Classic Bread | Dr. Schär | Italy | 3.49 | 300 |
| 130 | Maestro Vital Gluten-Free Cereal Bread | Dr. Schär | France | 4.31 | 350 |
| 131 | Gluten-Free Panini Bread Rolls | Dr. Schär | Spain | 2.83 | 75 |
| 132 | Gluten-Free Rustic Bread Flour | Dr. Schär | Spain | 5.68 | 1000 |
| 133 | Organic Gluten-Free Corn Bread with Sunflower Seeds | Schnitzer | Germany | 4.9 | 250 |
| 134 | Gluten-Free White Bread Slices | Dr. Schär | Spain | 2.87 | 300 |
| 135 | Gluten Free Panini Bread Rolls | Dr. Schär | Germany | 3.31 | 75 |
| 136 | Gluten-Free Rustico Bread | Dr. Schär | Hungary | 2.51 | 225 |
| 137 | Gluten-Free Sliced White Bread | Dr. Schär | Hungary | 2 | 200 |
| 138 | Gluten-Free Sliced Bread | Dr. Schär | Hungary | 3.58 | 240 |
| 139 | Gluten-Free Honey Corn Bread Mix | Williams-Sonoma | USA | 12.95 | 493.29 |
| 140 | Gluten Free Pumpkin Chocolate Chunk Quick Bread Mix | Williams-Sonoma | USA | 15.95 | 510.3 |
| 141 | Gluten-Free Multigrain Sour Dough Bread | Dr. Schär | Netherlands | 4.03 | 350 |
| 142 | Gluten-Free Homemade-Style Bread Slices | Dr. Schär | Italy | 5.05 | 240 |
| 143 | Gluten-Free Brown Rice Bread | Food For Life Baking | Mexico | 6.86 | 680 |
| 144 | Gluten-Free Sweet Bread Loaf | Dr. Schär | Italy | 5.11 | 370 |
| 145 | Gluten-Free Rustico Bread | Dr. Schär | Denmark | 4.58 | 225 |
| 146 | Gluten-Free Rice Almond Bread | Food For Life Baking | Mexico | 12.29 | 680.4 |
| 147 | Gluten-Free Wholegrain Oat Bread Mix | Bauck | Germany | | 500 |
| 148 | All Natural Gluten Free Rice Millet Bread | Food For Life Baking | Singapore | 9.37 | 680 |
| 149 | Gluten-Free Sliced White Bread | Genius Foods | France | 6.03 | 400 |
| 150 | Sesame Flavored Organic Gluten Free Bread | Schnitzer | Netherlands | 4.62 | 250 |
| 151 | Gluten-Free Bread with Cereals | Dr. Schär | France | 4.41 | 400 |
| 152 | Gluten Free White Bread | Dr. Schär | Brazil | 3.6 | 200 |
| 153 | Gluten Free Cereal Bread | Dr. Schär | Spain | 4.41 | 300 |

(continued)

**Table 9.2** (continued)

| Sr. | Product | Company | Market | Price | g |
|---|---|---|---|---|---|
| 154 | Organic Gluten Free Rustic Bread with Amaranth | Schnitzer | Germany | 6.01 | 250 |
| 155 | Easy Bake Gluten-Free Black Bread with Black Rice | Bauck | Spain | 9.36 | 475 |
| 156 | Gluten Free Happy Quinoa Bread | Arctis Tiefkühl-Backwaren | Germany | 6.83 | 600 |
| 157 | Gluten Free Sporty Seed-Coated Bread | Arctis Tiefkühl-Backwaren | Germany | 6.83 | 600 |
| 158 | Gluten Free Toast Bread | Arctis Tiefkühl-Backwaren | Germany | 5.96 | 600 |
| 159 | Gluten Free Sunny Sunflower Bread | Arctis Tiefkühl-Backwaren | Germany | 6.83 | 600 |
| 160 | Gluten Free Country Bread | Arctis Tiefkühl-Backwaren | Germany | 6.7 | 400 |
| 161 | Gluten Free Activity Bread | Arctis Tiefkühl-Backwaren | Germany | 7.1 | 400 |
| 162 | Gluten-Free Toasted Bread Slices | Dr. Schär | France | 5.85 | 83 |
| 163 | Gluten-Free Multigrain Rustic Bread | Dr. Schär | Brazil | 4.35 | 225 |
| 164 | Gluten-Free Piadina Flat Bread | Dr. Schär | Italy | 4.89 | 80 |
| 165 | Gluten-Free Classic Baker's Bread | Dr. Schär | France | 4.42 | 300 |
| 166 | Gluten Free Rose-Shaped Bread Rolls | Dr. Schär | Italy | 5.02 | 58 |
| 167 | Gluten Free Spiced Pecan Pumpkin Quick Bread Mix | Williams-Sonoma | USA | 14.95 | 510.3 |
| 168 | Organic Gluten-Free Golden Bread Rolls | Schnitzer | Finland | 4.78 | 125 |
| 169 | Organic Gluten-Free Sesame Bread | Schnitzer | Finland | 4.59 | 250 |
| 170 | Gluten-Free Multigrain Bread | Dr. Schär | Russia | 8.03 | 225 |
| 171 | Gluten-Free Brioche Bread | Dr. Schär | France | 5.17 | 370 |
| 172 | Gluten-Free Sliced Bread | Dr. Schär | Italy | 4.1 | 240 |
| 173 | Gluten Free White Bread | Country Life Bakery | Australia | 6.7 | 520 |
| 174 | Gluten-Free Multigrain Bread | Dr. Schär | Germany | 4.06 | 300 |
| 175 | Gluten-Free Oven Baked Bread | Dr. Schär | Italy | 6.15 | 300 |
| 176 | Quick and Light Gluten Free Bread Mix | Bauck | Germany | 6.27 | 475 |
| 177 | Naturally Gluten-Free Rustic Bread | Dr. Schär | France | 4.91 | 400 |
| 178 | Naturally Gluten-Free Buckwheat Bread Sticks | Dr. Schär | Finland | 5.76 | 50 |

(continued)

**Table 9.2** (continued)

| Sr. | Product | Company | Market | Price | g |
|---|---|---|---|---|---|
| 179 | Gluten Free White Bread | Country Life Bakery | Australia | 4.3 | 510 |
| 180 | Gluten Free White Sandwich Bread | Kinnikinnick Foods | USA | 5.7 | 567 |
| 181 | Gluten-Free Sandwich Bread | Dr. Schär | Austria | 5 | 400 |
| 182 | Wheat & Gluten Free Brown Rice Bread | Food For Life Baking | USA | 4.99 | 42.53 |
| 183 | Gluten Free Fruit Bread | Country Life Bakery | Australia | 4.66 | 510 |
| 184 | Gluten Free White Bread Buns | Dr. Schär | Hungary | 3.18 | 200 |
| 185 | Naturally Gluten-Free Bread | Dr. Schär | Austria | 3.5 | 400 |
| 186 | Gluten Free Bread Range | Unavailable | Netherlands | 3.6 | 400 |
| 187 | Gluten-Free Bread Rolls | Country Life Bakery | Australia | 3.39 | 360 |

**Table 9.3** List of gluten free pasta along with countries

| Sr. | Product | Company | Market | Price Dollar | g |
|---|---|---|---|---|---|
| 1 | 100% Natural Gluten-Free Cannelloni Pasta | Pol-Foods | Hungary | 3.17 | 500 |
| 2 | Organic & Gluten-Free Penne Pasta | Ancient Harvest | USA | 3.69 | 272.16 |
| 3 | Gluten Free Fusilli Pasta | Lidl | Sweden | 1.49 | 500 |
| 4 | Gluten Free Tagliatelle Pasta | REMA 1000 | Norway | 3.82 | 250 |
| 5 | Organic Gluten Free Fusilli Pasta | REMA 1000 | Norway | 3.28 | 500 |
| 6 | Organic Gluten Free Lentil Fusilli Pasta | REMA 1000 | Norway | 3.28 | 250 |
| 7 | Gluten Free Fusilli Pasta | Andriani | Tunisia | 1.63 | 340 |
| 8 | Organic Brown Rice Gluten Free Penne Pasta | San Remo | Indonesia | 3.02 | 250 |
| 9 | Organic Brown Rice Gluten Free Spaghetti Pasta | Pastificio Mennucci | Indonesia | 3.02 | 250 |
| 10 | Gluten Free Angel's Hair Pasta | Dr. SchŠr | Morocco | 3.37 | 250 |
| 11 | Gluten-Free Anellini Pasta | Dr. SchŠr | Slovenia | 2.35 | 250 |
| 12 | Gluten-Free Tagliatelle Pasta | Barilla | Norway | 3.83 | 300 |
| 13 | Gluten-Free Fusilli Pasta | Barilla | Norway | 3.91 | 400 |
| 14 | Organic Gluten Free Twist Pasta | Pates Grand'Mere | France | 2.42 | 250 |
| 15 | Original Gluten Free Fettuccine Pasta | San Remo Macaroni | Indonesia | 3.02 | 350 |
| 16 | Gluten Free Fusilli Pasta | Nachiogel foods for Al Mashreq | Egypt | 4.51 | 350 |

(continued)

**Table 9.3** (continued)

| Sr. | Product | Company | Market | Price Dollar | g |
|---|---|---|---|---|---|
| 17 | Gluten Free Chickpea Risoni Pasta | Alimentos El Dorado | Ecuador | 2.7 | 250 |
| 18 | Organic Gluten Free Penne Pasta | Dr. SchŠr | Austria | 3.13 | 350 |
| 19 | Gluten-Free Tortiglioni Pasta | Barilla/Barilla G. e R. Fratelli | Morocco | 5.01 | 400 |
| 20 | Gluten-Free Full Taste Spaghetti Pasta | Pastificio Lucio Garofalo | Portugal | 3.25 | 400 |
| 21 | Gluten Free Penne Rigate Pasta | Barilla/Barilla G. e R. Fratelli | Spain | 2.55 | 400 |
| 22 | Gluten Free Vegetable Mix Pasta Spirals | Healthy Generation | Ukraine | 1.65 | 300 |
| 23 | Organic Gluten-Free Buckwheat Maccheroni Pasta | Armida | Italy | 2.8 | 250 |
| 24 | Organic Gluten-Free Tortiglioni Multigrain Pasta with Quinoa | Sottolestelle | Italy | 2.86 | 340 |
| 25 | Organic Gluten-Free Multigrain Penne Pasta | Probios | Italy | 2.98 | 340 |
| 26 | Organic Gluten-Free Multigrain Fusilli Pasta | Probios | Italy | 2.98 | 340 |
| 27 | Organic Gluten-Free Multigrain Sedanini Pasta | Probios | Italy | 2.98 | 340 |
| 28 | Organic Gluten Free Brown Rice Pasta Fantasia | PGR Health Foods | Netherlands | 5.06 | 500 |
| 29 | Organic & Gluten-Free Rotini Pasta | Ancient Harvest | USA | 3.29 | 226.8 |
| 30 | Organic Gluten Free Red Lentils Penne Pasta | Ital Passion | France | 3.88 | 250 |
| 31 | Gluten-Free Spaghetti No. 5 Pasta | Pasta Berruto | Ethiopia | 4.05 | 400 |
| 32 | Gluten-Free Tri Colour Corn & Vegetable Pasta Shells | Aldi | Australia | 2.14 | 500 |
| 33 | Gluten Free Penne Pasta | Kaufland | Poland | 0.76 | 500 |
| 34 | Gluten Free Green Pea Pasta | Waitrose | UK | 2.56 | 250 |
| 35 | Gluten Free Penne Pasta | Bioalimenta | Germany | 3.88 | 500 |
| 36 | Gluten Free Corn and Rice Tagliatelle Pasta | Valle Roveto GlutenFree Srl | UK | 4.39 | 250 |
| 37 | Gluten Free Penne Pasta | Dr. SchŠr | Poland | 5.09 | 1000 |
| 38 | Gluten-Free Penne Rigate Pasta | Barilla/Barilla G. e R. Fratelli | Poland | 2.07 | 400 |
| 39 | Gluten Free Spaghetti Pasta | Fits Mandiri | Indonesia | 1.96 | 180 |
| 40 | Gluten-Free Cauliflower Linguini Pasta | Tribe 9 Foods | USA | 5.69 | 255.15 |

(continued)

**Table 9.3** (continued)

| Sr. | Product | Company | Market | Price Dollar | g |
|---|---|---|---|---|---|
| 41 | Gluten Free Corn Fusilli Pasta | Sam Mills | Mexico | 3.92 | 500 |
| 42 | Gluten-Free Penne Rigate Pasta No. 27 | Pastificio Riscossa | Kenya | 3.68 | 340 |
| 43 | Gluten-Free Ridged Penne Pasta | Barilla/Barilla G. e R. Fratelli | Chile | 3.58 | 400 |
| 44 | Gluten Free Spaghetti Pasta | Scamark | France | 1.52 | 400 |
| 45 | Gluten Free Penne Rigate Pasta No. 136 | Preferisco Foods | Canada | 3.56 | 500 |
| 46 | Organic Gluten-Free Cereal Penne Pasta | Dr. SchŠr | Spain | 2.8 | 350 |
| 47 | Gluten Free Rice, Corn and Quinoa Penne Pasta | Colombina | Colombia | 1.74 | 250 |
| 48 | Gluten-Free Spirali Pasta No. 50 | Pastificio R.F.M. | Egypt | 5.33 | 340 |
| 49 | Gluten Free Penne Pasta | Leader Price - DLP | Ivory Coast | 3.09 | 500 |
| 50 | Original Gluten Free Penne Pasta | San Remo Macaroni | Indonesia | 2.99 | 350 |
| 51 | Original Gluten Free Spaghetti Pasta | San Remo | Indonesia | 2.99 | 350 |
| 52 | Gluten-Free Quinoa Spirals Pasta | Orgran Health & Nutrition | Guatemala | 5.19 | 250 |
| 53 | Gluten Free Organic Brown Rice Pasta | Chacha Thai | Thailand | 1.51 | 225 |
| 54 | Gluten-Free Corn and Rice Lasagna Pasta | Molino di Ferro | Canada | 3.75 | 250 |
| 55 | Gluten-Free Ridged Penne Pasta | Barilla/Barilla G. e R. Fratelli | Italy | 1.65 | 400 |
| 56 | Gluten Free Macaroni Pasta | Etablissement Manseur.M | Algeria | 1.2 | 300 |
| 57 | Gluten-Free Spaghetti Pasta | Lidl | Hungary | 1.24 | 500 |
| 58 | Gluten Free Tortiglioni Buckwheat Pasta | Andriani | Sweden | 2.95 | 250 |
| 59 | Gluten Free Split Peas Fusilli Pasta | Biovence | France | 3.76 | 250 |
| 60 | Gluten Free Chickpea Spiral Pasta | Biovence | France | 3.56 | 250 |
| 61 | Gluten-Free Fusilli Pasta | Molinos R'o de la Plata | Argentina | 1.61 | 500 |
| 62 | Gluten-Free Penne Corn Pasta | Pol-Foods | Russia | 1.71 | 250 |
| 63 | Gluten Free Organic Red Rice & Chia Pasta | Perfect Earth Foods | UK | 4.99 | 225 |
| 64 | Gluten Free Anellini Pasta | Dr. SchŠr | Poland | 1.79 | 250 |

(continued)

**Table 9.3** (continued)

| Sr. | Product | Company | Market | Price Dollar | g |
|---|---|---|---|---|---|
| 65 | Gluten Free Fusilli Pasta | Dr. SchŠr | Poland | 1.36 | 250 |
| 66 | Gluten Free Spaghetti Pasta No. 104 | Preferisco Foods | Canada | 4.28 | 500 |
| 67 | Gluten Free Fusilli Pasta No. 1107 | Preferisco Foods | Canada | 4.28 | 500 |
| 68 | Gluten Free Cereal Penne Pasta | Dr. SchŠr | Poland | 3.38 | 350 |
| 69 | Gluten Free Pennoni Pasta with Legumes and Cereals | Storico Pastificio Garofalo | Italy | 3.24 | 400 |
| 70 | Gluten Free Fusilli Pasta with Legumes and Cereals | Storico Pastificio Garofalo | Italy | 3.24 | 400 |
| 71 | Gluten-Free Pea Pasta | Pasta d'Alba | Finland | 5.39 | 250 |
| 72 | Organic Gluten Free Buckwheat Penne Pasta | Organic Larder | UAE | 5.94 | 300 |
| 73 | Gluten-Free Gnocchetti Pasta with Spinach | Primaly | Italy | 6.24 | 400 |
| 74 | Gluten-Free Mini Pasta Shells | Molinos del Mundo | Peru | 4.02 | 400 |
| 75 | Penne Gluten Free Pasta with Cauliflower, Fava Beans & Rice Flour | Riviana Foods | USA | 2.49 | 283.5 |
| 76 | Gluten-Free Organic Buckwheat Penne Pasta | Probios | Germany | 2.78 | 250 |
| 77 | Gluten-Free Organic Rice Pasta | Probios | Germany | 2.78 | 400 |
| 78 | Gluten Free Egg Tagliatelle Pasta | Ocram | UK | 4.29 | 250 |
| 79 | Gluten-Free Penne Pasta | Dr. SchŠr | Netherlands | 3.56 | 500 |
| 80 | Gluten-Free Tagliatelle Pasta | Organico Realfoods | UK | 5.21 | 250 |
| 81 | Gluten Free Fusilli Pasta | Molino Andriani | Bangladesh | 7.67 | 400 |
| 82 | Gluten-Free Rice & Corn Penne Pasta | Riso Scotti | Brazil | 2.68 | 250 |
| 83 | Gluten-Free Wholegrain Rice Spaghetti #20 Pasta | Dialcos | Spain | 3.2 | 400 |
| 84 | Corn, Brown Rice and Quinoa Gluten-Free Spaghetti Pasta | Pastificio Lucio Garofalo | Brazil | 4.16 | 400 |
| 85 | Gluten-Free Caserecce Pasta | Arte & Pasta | Italy | 7.11 | 400 |
| 86 | Gluten Free Brown Rice, Quinoa & Cauliflower Fusilli Pasta | Kroger | USA | 2.79 | 226.8 |
| 87 | Gluten Free Red Lentil & Quinoa Fusilli Pasta | Kroger | USA | 2.79 | 226.8 |
| 88 | Gluten-Free Penne Pasta with Corn Flour | NORMA | Germany | 1.43 | 500 |

(continued)

**Table 9.3** (continued)

| Sr. | Product | Company | Market | Price Dollar | g |
|---|---|---|---|---|---|
| 89 | Original Gluten Free Penne Pasta | San Remo Macaroni | Thailand | 3.27 | 350 |
| 90 | Gluten Free Fusilli Pasta | Califood | Lebanon | 2.62 | 500 |
| 91 | Gluten-Free Dino Pasta | Morrisons | UK | 0.59 | 250 |
| 92 | Gluten-Free Penne Pasta | Scamark | France | 1.52 | 400 |
| 93 | Gluten-Free Corn and Rice Pasta | Valle Roveto GlutenFree Srl | UK | 4.56 | 250 |
| 94 | Gluten Free Mezze Penne Rigate No. 28 Pasta | Rummo | Hong Kong, China | 4.5 | 400 |
| 95 | Gluten Free Spaghetti No.2 Pasta | Pastificio Riscossa | Egypt | 5.38 | 400 |
| 96 | Gluten-Free Chickpeas Risoni Pasta | Alimentos El Dorado | Colombia | 2.72 | 250 |
| 97 | Gluten Free Fusilli Pasta | Coop Trading | Norway | 1.87 | 340 |
| 98 | Penne Rigate Gluten Free Pasta | Andriani | India | 2.81 | 400 |
| 99 | Gluten Free Brown Rice Spaghetti Pasta | Wegmans Food Markets | USA | 1.99 | 453.6 |
| 100 | Gluten Free Brown Rice Fusilli Pasta | Wegmans Food Markets | USA | 1.99 | 453.6 |
| 101 | Gluten Free Brown Rice Penne Pasta | Wegmans Food Markets | USA | 1.99 | 453.6 |
| 102 | Gluten Free Brown Rice Elbows Pasta | Wegmans Food Markets | USA | 1.99 | 453.6 |
| 103 | Gluten Free Brown Rice and Quinoa Spaghetti Pasta | Wegmans Food Markets | USA | 2.29 | 453.6 |
| 104 | Gluten Free Fusilli Pasta | Pol-Foods | Poland | 1.07 | 500 |
| 105 | Gluten Free Casarecce Pasta | Mamma Flora | UK | 7.04 | 400 |
| 106 | Gluten-Free Fusilli Pasta | Organico Realfoods | UK | 4.49 | 250 |
| 107 | Gluten Free Fusilli Pasta | Netto | Denmark | 1.48 | 500 |
| 108 | Gluten-Free Penne Rigate Pasta No. 27 | Pastificio Riscossa | Brazil | 3.58 | 340 |
| 109 | Gluten-Free Spaghetti No. 2 Pasta | Pastificio Riscossa | Brazil | 4.32 | 400 |
| 110 | Gluten Free Pasta with Cauliflower Penne, Fava Beans & Rice Flour | Riviana Foods | Puerto Rico | 2.68 | 283.5 |
| 111 | Organic Gluten Free Buckweat Fusilli Pasta | Alb-Gold Teigwaren | Poland | 2.61 | 250 |
| 112 | Gluten-Free Fusilli Pasta | Dr. SchŠr | France | 3.49 | 500 |
| 113 | Gluten-Free Egg Tagliatelle Pasta | Nutrition & SantŽ | France | 4.14 | 250 |
| 114 | Gluten-Free Penne Pasta | Aldi | Hungary | 3.37 | 500 |

(continued)

**Table 9.3** (continued)

| Sr. | Product | Company | Market | Price Dollar | g |
|---|---|---|---|---|---|
| 115 | Gluten Free Penne Pasta | Pol-Foods | France | 1.31 | 500 |
| 116 | Corn and Rice Gluten-Free Caserecce no.37 Pasta | Dialcos | Poland | 3.23 | 400 |
| 117 | Gluten Free Tagliatelle Pasta | Bezgluten | Poland | 1.05 | 250 |
| 118 | Gluten Free Tortellini Pasta Filled with Raw Ham | Dr. SchŠr | Switzerland | 4 | 250 |
| 119 | Gluten Free Spaghetti Pasta | Pol-Foods | UK | 1.26 | 500 |
| 120 | Gluten Free Penne Rigate Pasta | Barilla/Barilla G. e R. Fratelli | Hungary | 2.97 | 400 |
| 121 | Organic Gluten-Free Green Pea Penne Pasta | Fabijanski | Poland | 3.09 | 250 |
| 122 | Gluten-Free Elbow Pasta | HammermŸhle | Germany | 2.61 | 500 |
| 123 | Gluten-Free Brown Rice Fusilli Pasta | Eurospital | Italy | 3.21 | 250 |
| 124 | Gluten Free Fusilli Pasta | Lidl | France | 1.42 | 500 |
| 125 | Gluten-Free Spaghetti Pasta | Organico Realfoods | UK | 4.85 | 250 |
| 126 | Gluten-Free Alphabet Shaped Pumpkin Pasta | Pol-Foods | Germany | 2.52 | 400 |
| 127 | Gluten-Free Beetroot Pasta Animals | Pol-Foods | Germany | 2.52 | 400 |
| 128 | Gluten-Free Bucatini Pasta | Pol-Foods | Germany | 2.19 | 400 |
| 129 | Gluten-Free Alphabet Shaped Tomato Pasta | Pol-Foods | Germany | 2.52 | 400 |
| 130 | Gluten-Free Animal Shaped Tomato Pasta | Pol-Foods | Germany | 2.52 | 400 |
| 131 | Gluten-Free Animal Shaped Pumpkin Pasta | Pol-Foods | Germany | 2.52 | 400 |
| 132 | Gluten Free Penne Pasta | Bioalimenta | Germany | 3.91 | 500 |
| 133 | Gluten-Free Organic Brown Rice Farfalle Pasta | Jovial Foods | USA | 4.19 | 340.2 |
| 134 | Wild Mushroom Gluten-Free Pasta Sauce | Dave's Gourmet | USA | 11.45 | 722.93 |
| 135 | Gluten-Free Maccheroni Rigati Pasta | La Fabbrica della Pasta di Gragnano | Italy | 3.59 | 500 |
| 136 | Gluten-Free Gems Pasta | BiAglut | Italy | 3.01 | 250 |
| 137 | Gluten Free Fusilli Pasta | ICA | Sweden | 1.54 | 500 |
| 138 | Gluten Free Casarecce Pasta | ICA | Sweden | 1.65 | 500 |
| 139 | Gluten Free Konjac Flour Tagliatelle Pasta | Calorie Watchers | Poland | 2 | 385 |
| 140 | Gluten Free Fusilli Pasta | Pasta Lensi | UK | 2.17 | 400 |
| 141 | Organic Gluten Free Rice Gemelli Pasta | Fabijanski | Poland | 1.03 | 225 |

(continued)

**Table 9.3** (continued)

| Sr. | Product | Company | Market | Price Dollar | g |
|---|---|---|---|---|---|
| 142 | Gluten-Free Lasagne Pasta | Barilla/Barilla G. e R. Fratelli | Germany | 3.33 | 250 |
| 143 | Corn, Brown Rice & Quinoa Gluten Free Casarecce Pasta | Pastificio Lucio Garofalo | Germany | 4.43 | 400 |
| 144 | Gluten-Free Animal Shaped Pasta | Pol-Foods | Germany | 2.21 | 400 |
| 145 | Gluten Free Green Pea Penne Pasta | Sam Mills | Germany | 3.13 | 250 |
| 146 | Gluten-Free Red Lentil Penne Pasta | Sam Mills | Germany | 3.13 | 250 |
| 147 | Gluten-Free Black Bean Penne Pasta | Sam Mills | Germany | 3.13 | 250 |
| 148 | Gluten-Free Fusilli Pasta | DM Drogerie Markt | Czech Republic | 2.63 | 400 |
| 149 | Gluten Free Rice, Corn and Quinoa Spaghetti Pasta | Colombina | Colombia | 2.07 | 250 |
| 150 | Gluten Free Rice, Corn and Quinoa Fusilli Pasta | Colombina | Colombia | 1.93 | 250 |
| 151 | Gluten-Free Tagliatelle Pasta | Barilla/Barilla G. e R. Fratelli | France | 2.68 | 300 |
| 152 | Organic Gluten Free Artisan Pasta Red Lentil Tagliatelle | Peregrine Trading | UK | 6.2 | 250 |
| 153 | Gluten-Free Tagliatelle Pasta | Nutrition & SantŽ | UAE | 9.99 | 250 |
| 154 | Gluten-Free Vegetable Rice Spiral Pasta | Orgran Health & Nutrition | Netherlands | 3.59 | 250 |
| 155 | Gluten-Free Tagliatelle Pasta | Bioalimenta | Brazil | 4.36 | 250 |
| 156 | Gluten Free Spaghetti Pasta No. 5 | Dr. SchŠr | Argentina | 3.02 | 250 |
| 157 | Gluten Free Fusilli Pasta No. 24 | Dr. SchŠr | Argentina | 3.02 | 250 |
| 158 | Gluten Free Penne Pasta No. 21 | Dr. SchŠr | Argentina | 3.02 | 250 |
| 159 | Gluten-Free Maccheroni Pasta | Semper | Finland | 3.86 | 500 |
| 160 | Gluten-Free Rotini Pasta | Free to Eat | Panama | 4.99 | 227 |
| 161 | Gluten Free Fusilli Pasta | Pastificio Attilio Mastromauro Granoro | India | 4.32 | 400 |
| 162 | Gluten Free Fusilli Pasta | DM Drogerie Markt | Germany | 2.66 | 400 |
| 163 | Gluten-Free Angel Hair Pasta | Rainbow Mountain | Peru | 4.19 | 250 |
| 164 | Gluten-Free Chia Spaghetti Pasta | Alimentos El Dorado | Peru | 4.07 | 250 |
| 165 | Gluten-Free Rice Spaghetti Pasta | Alimentos El Dorado | Peru | 2.99 | 250 |
| 166 | Gluten-Free Royal Quinoa Spaghetti Pasta | Alimentos El Dorado | Peru | 4.07 | 250 |

(continued)

**Table 9.3** (continued)

| Sr. | Product | Company | Market | Price Dollar | g |
|---|---|---|---|---|---|
| 167 | Gluten-Free Corn Spaghetti Pasta | Alimentos El Dorado | Peru | 2.99 | 250 |
| 168 | Gluten-Free Fusilli Pasta | Scamark | France | 1.69 | 400 |
| 169 | Gluten-Free Royal Quinoa Fusilli Pasta | Alimentos El Dorado | Argentina | 3.11 | 250 |
| 170 | Gluten Free Tagliatelle Nests Egg Pasta | San Remo | Australia | 3.46 | 250 |
| 171 | Gluten Free Organic Chia Pasta | Chacha Thai | Australia | 4.11 | 225 |
| 172 | Gluten Free Organic Chia Pasta | Chacha Thai | Australia | 4.11 | 225 |
| 173 | Gluten-Free Penne Pasta | Kaufland | Croatia | 1.95 | 500 |
| 174 | Gluten-Free Penne Rigate Pasta | Molino di Ferro | Netherlands | 2.4 | 250 |
| 175 | Penne Rigate Gluten Free Pasta | Pastificio Antonio Pallante | Estonia | 3.52 | 400 |
| 176 | Gluten Free Gnocchetti Pasta | Organico Realfoods | UK | 4.45 | 250 |
| 177 | Gluten-Free Ricotta & Spinach Tortelli Pasta | Bofrost | Austria | 12.3 | 500 |
| 178 | Gluten Free Quinoa Spirals Pasta | Orgran Health & Nutrition | Australia | 2 | 250 |
| 179 | Gluten Free Organic Brown Rice & Chia Pasta | Chacha Thai | Australia | 4.21 | 225 |
| 180 | Gluten Free Organic Red Rice & Chia Pasta | Chacha Thai | Australia | 4.21 | 225 |
| 181 | Organic Gluten Free Spirelli Rice Pasta | Naturata | Spain | 3.14 | 250 |
| 182 | Gluten-Free Corn Spaghetti Pasta | Lidl | Poland | 0.78 | 500 |
| 183 | Gluten-Free Corn Fusilli Pasta | Lidl | Poland | 0.78 | 500 |
| 184 | Gluten-Free Tortiglioni Pasta | Barilla/Barilla G. e R. Fratelli | Greece | 2.44 | 400 |
| 185 | Organic Gluten Free Artisan Rice & Quinoa Tagliatelle Pasta | Peregrine Trading | UK | 6.19 | 250 |
| 186 | Gluten Free Organic Rice Pad Thai Pasta | Chacha Thai | Thailand | 1.54 | 225 |
| 187 | Bread, Pizza, Pasta and Cakes Gluten-Free Baking Mix | Lidl | Italy | 2 | 500 |
| 188 | Gluten Free Fusilli Pasta | Kroger | USA | 2.99 | 340.2 |
| 189 | Gluten-Free Rice Pasta Spirals | Orgran Health & Nutrition | Australia | 2.79 | 250 |
| 190 | Gluten-Free Fusilli Pasta | Lidl | Czech Republic | 1.32 | 500 |
| 191 | Gluten-Free Chia Fusilli Pasta | Alimentos El Dorado | Argentina | 2.96 | 250 |

**Fig. 9.3** Global gluten free pasta market by region

# References


Amanda topper, Non-celiacs drive gluten-free market. (2017). https://www.mintel.com/press-centre/food-and-drink/gluten-free-foods-surge-63-percent

Foschia, M., Horstmann, S., Arendt, E. K., & Zannini, E. (2016). Nutritional therapy–Facing the gap between coeliac disease and gluten-free food. *International Journal of Food Microbiology, 239*, 113–124.

Gallagher, E. (Ed.). (2009). *Gluten-free food science and technology*. Wiley.

Gluten Free Food Market Size and Share Evolution to 2021 by Growth Insight. (2021). *Key development, trends and forecast 2027*. https://www.marketwatch.com/press-release/gluten-free-food-market-size-and-share-evolution-to-2021-by-growth-insight-key-development-trends-and-forecast-2027-2021-01-18

Grandview research, Gluten-Free Products Market Size, Share & Trends Analysis Report By Product (Bakery Products, Dairy/Dairy Alternatives), By Distribution Channel (Grocery Stores, Mass Merchandiser), By Region, And Segment Forecasts, 2020–2027. https://www.grandviewresearch.com/industry-analysis/gluten-free-products-market

Miranda, J., et al. (2014). Nutritional differences between a gluten-free diet and a diet containing equivalent products with gluten. *Plant Foods for Human Nutrition, 69*(2), 182–187.

Rosell, C. M., & Matos, M. E. (2015). Market and nutrition issues of gluten-free foodstuff. Advances in the understanding of gluten related pathology and the evolution of gluten-free foods, 675–713.

# Index






N. Singh Deora et al. (eds.), *Challenges and Potential Solutions in Gluten Free Product Development*, Food Engineering Series,
https://doi.org/10.1007/978-3-030-88697-4